I0464255

Debating Diagnosis:
Perspectives on the DSM 5

Debating Diagnosis:
Perspectives on the DSM 5

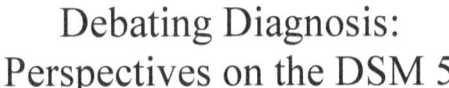

Nora L. Ishibashi, Ph.D., Editor

© 2015 Nora L. Ishibashi, Ph.D.
405 North Wabash Avenue, Suite 1303
Chicago, IL 60611

All rights reserved. This book or any portion thereof may not be reproduced or used in any manner whatsoever without the express written permission of the publisher except for the use of brief quotations in a book review or scholarly journal.

First Printing: 2015

ISBN: 978-1-312-79697-3

Ordering information:

This book may be ordered from http://www.lulu.com

Dedication

This book is dedicated to all caregiving professionals who have committed themselves to lives of care and responsibility.

Contents

Preface

Nora L. Ishibashi, Ph.D.

The *Diagnostic and Statistical Manual of Mental Disorders* (DSM) is published by the American Psychiatric Association as a standard tool used by most mental health practitioners in the United States to provide diagnoses for psychological distress. Like earlier versions of the DSM, the current edition, which is the fifth complete revision, continues to create controversy as experts argue the merits and pitfalls of its definitions and implications.

This book collects papers from master's degree students in the Social Work School at Loyola University of Chicago. As a part of their class in Human Behavior in the Social Environment, students study the DSM as one of the many ways social workers assess the needs of their clients. These papers represent their thoughts on assessment and diagnosis, including wider views and theoretical perspectives.

Psychological distress is difficult to assess because it is an assessment of the subjective experience of a client through the subjective experience of the social worker. Unlike medical problems, psychological problems cannot be detected using blood tests, EEG's, blood pressure monitors, or other mechanical measures. We are left with our judgment and the client's description of his or her experience as means of understanding the problem. As a result, assessment includes both interpretation and description, which means naming the problem can be said to identify a reality but can also be said to create a reality.

Our work in joining with people to improve their lives is a profound responsibility with the potential for harm as well as help. We influence people in ways we can see but also in ways we cannot see. Social workers have always been aware of the potential impact we have on our clients, and the social work profession has over its long tradition generated many thoughtful books, much careful research, and countless knowledge-building conferences, meetings, and curriculums. We have taken our responsibilities seriously and developed ethical standards and caregiving approaches with the client's well-being as our central interest.

These papers represent the efforts of the current cohort of social work students to come to grips with the skills needed to assess the distress clients will bring to them. Social work is a noble profession, and students who are embarking on this path are dedicated, caring, and courageous. They engage with current theories, ideas, and practices and measure them against their own life experience, their opinions about human nature, and their aspirations to change the world, mostly one person at a time. You will find these papers to be thoughtful and varied. They represent the best of a new generation of social workers.

Autism Spectrum Disorder and the DSM V

Elizabeth Broder-Oldach

The diagnostic criteria for Autism Spectrum Disorder (ASD) in the 5th edition of the Diagnostic and Statistics Manual (DSM V) include: "…persistent impairment in reciprocal social communication and social interaction (Criterion A), and restricted, repetitive patterns of behavior, interests, or activities (Criterion B)" (DSM V, 2013, p. 53). For Criterion A, deficits in communication and social interaction can include difficulties with understanding relationships, non-verbal communication, and social-emotional reciprocity (DSM V, 2013). Deficits in Criterion B can include inflexibility or insistence on sameness; fixations on particular, often unusual, interests; and repetitive movements, speech, or use of objects (DSM V, 2013). Sensory over- or under- sensitivity can also be part of Criterion B (DSM V, 2013). Additionally, as a neurodevelopmental disorder, the symptoms of ASD must be present from early childhood, though the DSM V notes that the presentation of symptoms can be delayed depending on cultural context and the social-emotional demands made of the child (DSM V, 2013). An intellectual disability or language impairment can, but does not always, accompany ASD (DSM V, 2013). As the name "Autism Spectrum Disorder" implies, people with ASD have a wide range of abilities and needs for support. This is perhaps why there is so much disagreement about what autism is and how people with ASD should be treated.

The diagnosis of ASD seems like it should be very straightforward. A person has symptoms that meet the criteria for ASD and so they receive the diagnosis because, presumably, they *have* the disorder. This explanation supposes that the DSM is filled with disorders that can be empirically proven. In reality, the purpose of the DSM is to categorize clusters of symptoms and identify them as a disorder. In many cases, there is no empirical way to diagnose the disorder. Furthermore, diagnoses are not static or absolute. An individual who once was diagnosed with Pervasive Developmental Disorder – Not Otherwise Specified (PDD-NOS) or Asperger's syndrome under the DSM IV would now receive the diagnosis of Autism Spectrum Disorder (DSM V, 2013). It is unsettling that a diagnosis that has become a part of many people's identity can simply be changed. This knowledge makes one wonder, what does a diagnosis of ASD really mean? It turns out that it can mean many, many different things.

ASD and the Theory of Neurodiversity

According to Cascio (2012) there are three main perspectives on what Autism Spectrum Disorder is and what it means. The first is a disease model; ASD is a disease that can eventually be cured. Those who align with this school of thought are often parents of children who are significantly impacted by autism. This group often advocates that funding be allocated to finding a cure for autism. The second perspective is a disability model; ASD is a disability with symptoms that can be managed, with support, across the lifespan. The third model theorizes that ASD is a form of neurodiversity or, "a naturally occurring, and even positive [variation] in human cognitive wiring that should be celebrated rather than eliminated" (Cascio, 2012, p. 274). Proponents of these last two models often advocate that funding for services for those with ASD should be prioritized over funding dedicated to finding a cure.

The idea of autism as neurodiversity opens up the diagnosis of ASD to allow for an interpretation more based in self-determination. For example, communication challenges are one of the diagnostic criteria for ASD. However, according to Jurecic (2007), "some students with Autism Spectrum Disorder may not want to express themselves more conventionally. A growing and vocal set of autistic activists - under the banner of 'neurodiversity' - are demanding that autism be accepted and respected not as a disorder, but as a variation on 'brain wiring" (p. 423). Similarly, there is a split regarding the preferred, dignifying language for referring to a person with ASD. Generally, the disability community advocates for person first language, such as "person with autism" (Snow, 2013). Naming the person before the disability allows the speaker to describe the diagnosis as a characteristic of a complex and multi-faceted person rather than an overly defining feature. However, members of the neurodiversity movement prefer to be referred to with identity first language, in this case, "autistic person" (Brown, n.d.). The neurodiversity model embraces the label "autism" as a positive descriptor for people with ASD where the disease model makes "autism" a dirty word and something to fight, like cancer. This seemingly simple disagreement about language underscores the differing opinions regarding how to think about ASD. Likewise, it emphasizes the importance of taking the individual's experiences and preferences into the diagnostic process, something which the DSM V does very little of.

The DSM V does not comment on whether ASD is a disease or a disability. However, it does pathologize ASD in a way that will not allow for the neurodiversity perspective, environmental factors, or the individual experience of the person with ASD to truly be taken into account. It is important to evaluate the DSM V's construction of ASD through the lens of the theory of neurodiversity. As previously mentioned, the DSM V does not take into account the perspective of people with autism. In the DSM, neurological differences are treated as pathology and symptoms to be treated, sometimes without the consent of the individual who has ASD. Neurodiversity, on the other hand states that, "neurological variation is not only natural, but is central to the success of the human species" (McGee, 2012, p. 12). It is through the perspective of neurodiversity that one can start to see the impact that individual experiences and preferences, cultural norms, and environmental factors have on diagnosis of ASD.

The Words of the DSM V: Who has the Disability?

As previously mentioned, the DSM V uses judgmental language that assumes that the deficit lies in the individual with ASD rather than in the interaction of the individual with his or her environment, or even the environment itself. The DSM V states:

> The stage at which functional impairment becomes obvious will vary according to characteristics of the individual and his or her environment. Core diagnostic features are evident in the developmental period, but intervention, compensation, and current supports may mask difficulties in at least some contexts (DSM V, 2013, p. 53.)

The DSM V does state that the functional impairment of the individual will vary depending on his or her environment but then says that intervention and the environment, in the form of supports, can only *mask* difficulties. This suggests that the deficits are inherent in the individual. However, if environment can alleviate impairment then perhaps the source of the impairment is the environment, not the individual. The field of disability studies has been advocating for this position for many years. According to Lewiecki-Wilson et al. (2008):

> Disability studies holds that mainstream culture often behaves in an *ablest* way: assuming that disability is inherently bad, that a disability is a deficit justifying intolerance and stigma, that it should be cured or overcome; assuming that people with disabilities can be spoken and acted for; and allowing individuals to make these assumptions by claiming a position as ultimately *non-disabled* and therefore unmarked and entitled to

diagnose and stigmatize others. Ableist positioning is thus *normative*....[Disability studies] argues that disability is a social construction." (p. 314-315).

As with the perspective of neurodiversity, disability studies takes a broader view of societally perceived impairments and reinterprets them as differences. Lewiecki-Wilson et al. (2008) go on to describe that though disability is a social construction, it does not mean that the disability does not exist or that the person with a disability does not experience real impairment or challenges. However, rather than explaining the deficit as inherent in the person, disability studies argues that the impairment comes from the way society constructs the idea of functioning. Similarly, the neurodiversity perspective does not necessarily disregard the need for services and support for those with ASD as long as the perspective of the person receiving the services is considered.

Orsini writes, "[i]t is possible to support funding for autism care and support, recognizing that some opponents view the behavioral interventions championed by autism advocates as an assault on autistic personhood" (Orsini, 2012, p. 808). According to Orsini (2012), the neurodiversity movement embraces some of the characteristics or "symptoms" of ASD as part of a person's identity. For example, author John Elder Robison discusses his experience building guitars for the band, KISS, that would shoot off fireworks. He had no training in this area but rather attributes his ability to focus intensely and to develop inventions like the guitars to being on the autism spectrum (Robison, 2007).

The neurodiversity perspective also brings in a much-needed reminder about self-determination and autonomy when it comes to treatment. This perspective integrates the characteristics of ASD along with a person's lived experience advocating that, whenever possible, the person with ASD should have a say in their own treatment and even diagnosis. Carly Fleischmann (n.d.) expresses a similar sentiment on her website. Fleischmann is a young woman with ASD. When Fleischmann was young, her parents believed that she had significant intellectual impairment and could not communicate (Fleischmann, n.d.). At age ten, Fleischmann learned how to communicate by typing, allowing others to understand her inner experience. On her website, Fleischmann talks about including people with ASD in the dialogue about diagnosis, experience, and treatment. She says,

> I think people get a lot of their information from so-called experts but I think what happens is that experts can't give an explanation to certain questions. How can you explain something you have not lived or if you don't know what it's like to have it? If a horse is sick, you don't ask a fish what's wrong with the horse. You go right to the horse's mouth. (Fleischmann, n.d.)

The Words of the DSM V: What is Impairment?

One of the criteria for ASD, and indeed any disorder in the DSM V, is that "the features [of the disorder] must cause clinically significant impairment in social, occupational, or other important areas of current functioning" (DSM V, 2013, p. 55). The definition of impairment, in this case, comes from socially constructed ideology about success and appropriate behavior. In fact, what causes impairment is not the disorder, but rather society's response to it. People without autism, or "neurotypicals" (NT) as the neurodiversity perspective calls them, find the behaviors of people with ASD undesirable and so they are deemed pathological, something to be eliminated (Leweicki-Wilson et al., 2008). Perhaps some behaviors of people with ASD should be eliminated; head banging and other self-injurious behaviors come to mind. However, rather than calling these behaviors weird, undesirable, or dangerous it is imperative to look at the need these behaviors are meeting. There is a need to ask, what is the experience of the individual? Which characteristics does the individual want to keep?

There is an additional conundrum in that many people with ASD do not communicate in a way that practitioners and parents are able to understand. Because of this, currently, it is not always possible to give people with ASD a voice at the table in determining which behaviors do impair their quality of life and which they consider part of their identity. Similarly, it is not always possible to determine which impairments are insurmountable and which practitioners just do not have a solution for. For example, there have been many examples of people with ASD who were believed to have significant intellectual impairments when in reality they just did not have a functional means of communication (Fleischmann, 2014). These are both issues that need to continue to be addressed. It is not acceptable for professionals to give up on the idea of self-determination simply because they have not yet figured out how to communicate effectively with a client.

In this area, there is also a tension between NT parents and practitioners and the people who have a diagnosis of ASD, both in terms of what traits are distressing or impairing and also which traits practitioners should strive to eliminate or cure. According to Bagatell (2010) some "[NT] parents see autism as a condition that prevents their child from living a happy and productive life and causes their child to suffer physically and emotionally. Autism is seen to mask the 'real child' rather than constituting an essential part of who the child is" (p. 45). Bagatell (2010) goes on to compare ASD to other behaviors or traits that practitioners have judged as negative and sought to eliminate such as left-handedness and homosexuality. Bagatell (2010) cautions that practitioners have historically made errors in deciding which behaviors are pathological. These decisions have come at a high cost to those with those traits or behaviors. It is the responsibility of practitioners to find a functional way to communicate with people with ASD, especially given stories like Fleischmann's. Otherwise, there is no way to know if treatment and even diagnosis is doing more harm than good.

The Words of the DSM V: Disruptive to Who?

Perhaps the most judgmental statement the DSM V makes regarding those diagnosed with ASD is regarding their "disruptive or challenging behaviors." The DSM V states, "...disruptive/challenging behaviors are more common in children and adolescents with autism spectrum disorder than other disorders, including intellectual disability" (DSM V, p. 55). This is a description of the perception that clinicians, teachers, and loved ones may have of the behavior of the person with ASD. People with ASD engage in a range of behaviors that may be unusual for NT clinicians, family members, and community members to see. Some behaviors include flapping hands, repeating heard words or syllables, organizing rather than playing with toys, head banging, and self-injury (DSM V, 2013). However, the measure of how disruptive or challenging a behavior is, is subjective and dependent on many factors including the caregiver's mood, the environment the behavior takes place in, and the understood cause of the behavior. Furthermore, the DSM V neglects to explain why the person with ASD might have more disruptive behavior. A statement like this warrants an exploration of whether this disruptive behavior is a result of how the person with ASD is treated, a mismatch between the environment and person, or something else entirely.

Mayer Shevin (1987) addresses this tension between behavior and intent in his poem, "The Language of Us and Them." Shevin elegantly makes the point that many of the traits society takes for granted as simply "human" are misunderstood when seen in slightly different forms coming from a person with a disability. The poem points out that behavior is often interpreted from the point of view of the person observing rather than experiencing the behavior. In a segment of his poem, Shevin (1987) writes:

> We like things
>> They fixate on objects
>
> We try to make friends
>> They display attention seeking behavior
>
> We take breaks
>> They display off task behavior
>
> We stand up for ourselves
>> They are non-compliant…
>
> We are human
>> They are…?

The poem reflects the theory of neurodiversity, questioning why the behaviors of a person with ASD and without ASD cannot both be examined from the perspective of that person's lived experience. A parody website created by an individual with ASD also addresses the nature of judgment from the NT population and reflects the judgmental nature of the DSM V as well. The website creates a fictional "Neurotypical Syndrome" defining it as: "a neurobiological disorder characterized by preoccupation with social concerns, delusions of superiority, and obsession with conformity" (Muskie, 2002). The DSM V makes the mistake of interpreting behavior and judging it based on observation rather than including the experience of those living with ASD.

The Words of the DSM V: The Prognosis

The DSM V goes on to paint a picture of a bleak future for those with ASD, one that lacks real understanding that people with ASD are even human.

Only a minority of individuals with autism spectrum disorder live and work independently in adulthood; those who do tend to have superior language and intellectual abilities and are able to find a niche that matches their special interests and skills. In general, individuals with lower levels of impairment may be able to function independently. However, even these individuals may remain socially naïve and vulnerable, have difficulties organizing practical demands without aid, and are prone to anxiety and depression. Many adults report using compensation strategies and coping mechanisms to mask their difficulties in public but suffer from stress and effort of maintaining a socially acceptable façade. (DSM V, 2013 p. 56)

Again, the DSM V describes some people with ASD as being able to "mask" their difficulties. While some people with ASD may experience stress or pressure from essentially communicating from a different cultural perspective, the idea of "masking" difficulties in an effort to maintain a "socially acceptable façade" is a disturbing one. A more respectful and dignifying perspective would argue that a person with ASD's way of being and communicating is simply a difference that might cause challenges when interacting with the NT population. Instead the DSM talks about those with ASD as if they were trying to pass as human and not quite making the mark.

Another dimension that this passage raises but does not adequately explore is the question of cultural differences. The DSM does acknowledge that, "cultural differences will exist in norms for social interaction, nonverbal communication, and relationships, but individuals with autism spectrum disorder are markedly impaired against the norms for their cultural context" (DSM V, 2013, p. 57). However, the DSM does not discuss the idea that the level of impairment experienced by those with ASD is cultural too. The disability culture has a difference of opinion when it comes to where the impairment lies. Could the impairment be a result of a lack of creative solutions in society? How do differences in cultural goals around independent living impact how a person's disability is experienced and viewed? The point of view that few people with ASD can live independently shows bias and a lack of creativity in looking at how communities can be better organized to meet the needs of people with ASD.

Stigmatization of ASD the DSM V

Over and over again, the DSM V presents ASD in a highly judgmental and stigmatized way. This is consistent with the way that the media presents ASD as well. As Holton, Farrell, and Fudge (2014) explain, the media presents ASD "as a shameful, isolating, and burdensome impairment that disrupts the lives of those diagnosed with it, their friends and families, and the communities they live in" (p. 191). The DSM and the media could stand to listen more to those with ASD who can speak to the experience firsthand. John Elder Robison, an often-published author with ASD says, "…we do not like hearing that we are defective or diseased. We do not like hearing that we are part of an epidemic. We are not problems for our parents, or society, or genes to be eliminated. We are people" (Joss, 2013). Fleischmann (2014) expresses similar sentiments, "In a society that depends on words for a person to be apart [sic] of it, it is hard to be a person that doesn't have words." The way that society values traditional communication often leaves people with ASD unable to share their experience and their thoughts. It is apparent that the language of the DSM V continues to feed into the idea of people with ASD being thought of as "less than." Words reflect the attitudes of those who wrote them and it seems that the authors of the DSM V do not hold the ASD community in high regard. Considering that the purpose of the DSM is to help serve and provide supports for people with ASD this is a grave injustice. There is a need to use more respectful and dignifying language in a manual that is used widely by clinicians serving this population. There is a need to consider the experience of having ASD from the perspective of those diagnosed with it.

Conclusion

People with the diagnosis of ASD vary widely in their experiences and also their capabilities in the context of the way society is structured. There is a need to examine all these perspectives on the diagnosis of ASD. Particular attention needs to be given to the way that practitioners and caregivers understand the experience of those who have been diagnosed with ASD, many of whom do not communicate traditionally.

The real question is, does ASD even belong in the DSM V? Is autism a disorder or, as the neurodiversity perspective describes it, neurological diversity to be "celebrated rather than eliminated" (Cascio, 2012, p. 274). The symptoms that lead to a diagnosis of ASD may require support given the way our society is currently structured and so, people with ASD should be entitled to services. Currently, these supports are only accessible with a diagnosis. However, this requirement is problematic in many ways. The diagnosis should make the challenges that a person with ASD faces better, not worse. The diagnosis should not burden a person with the expectation that their behavior will be "disruptive or challenging" to others. The diagnosis should be useful and respectful. Unfortunately, for many, the diagnosis of autism leads to stigma and judgment. Because of all this, a different view of society's perceptions of ASD is needed, starting with the way that the DSM V talks about people with ASD. This will hopefully ensure that people with ASD are treated with respect both in society and in the DSM.

References

American Psychiatric Association. (2013). *Diagnostic and statistical manual of mental disorders: DSM-5*. Washington, D.C: American Psychiatric Association.

Bagatell, N. (2010) From cure to community: Transforming notions of autism. *Journal of the Society for Psychological Anthropology*. 38(1). 33-55.

Brown, L. (n.d.). Identity-first language. *Autistic Self Advocacy Network*. Retrieved from http://autisticadvocacy.org/identity-first-language/.

Cascio, M.A. (2012) Neurodiversity: Autism pride among mothers of children with autism spectrum disorders.. *Intellectual and Developmental Disabilities*, 50(30), 273-283.

Fleischmann, C. (2014). Carly's voice. *Twitter*. Retrieved from https://twitter.com/CarlysVoice.

Fleischmann, C. (n.d.) Carly's voice. *Carly's Voice*. Retrieved from http://carlysvoice.com/home/aboutcarly/.

Holton, A., Farrell, L., Fudge, J. (2014) A threatening space?: Stigmatization and the framing of autism in the news. *Communication Studies*. 65(2). 189-207.

Joss, L. (2013). John Elder Robison leaves Autism Speaks amid controversy. *Autism Daily Newscast*. Retrieved from http://www.autismdailynewscast.com/john-elder-robison-leaves-autism-speaks-amid-controversy/4773/laurel-joss/.

Jurecic, A. (2007). Neurodiversity. *College English*. 69(5). 421-442.

Lewiecki-Wilson, C., Dolmage, J., Heilker, P., Jurecic, A. (2008). Two comments on "neurodiversity." *College English*. 70(3). 314-325.

McGee, M. (2012). Neurodiversity. *Contexts*. 11(3). 12-13.

McLeod, S. (2007). Humanism. *Simply Psychology*. Retrieved from http://www.simplypsychology.org/humanistic.html.

Muskie. (2002). What is NT. *Institute for the study of the Neurologically Typical*. Retrieved from http://isnt.autistics.org/

Orsini, M. (2012). Autism, neurodiversity and the welfare state: The challenges of accommodation neurological differences. *Canadian Journal of Political Science*. 45(4). 805-827.

Robison, J.E. (2007). *Look me in the eye: My life with Asperger's.* New York: Crown Publishers.

Shevin, M. (1987). The language of us/them. *Inspirational Articles and Poems.* Retrieved from: http://www.oafccd.com/lanark/poems/language.html

Snow, K. (2013). People first language. *Disability is Natural.* Retrieved from http://www.disabilityisnatural.com/images/PDF/pfl09.pdf.

DSM-5 and Psychodynamic Theory: Looking at Depression

Renata Calderon

Depression is a disorder that many suffer from, and from a professional and clinical point of view, the DSM-5 is the go-to tool in order to assess and diagnose an individual in the United States. For social workers, the symptoms alone do not provide much in terms of assessing the client as a whole while looking at person-in-environment. There are more factors than just the symptoms alone that clinicians look at when assessing an individual. Social Work clinicians make use of a biopsychosocial assessment, which encompasses several aspects of an individual's life, with symptoms only being a piece that makes up the puzzle. Essentially the DSM-5 gives the problem a name without providing the cause. The DSM-5 does provide a foundation for a starting point, and when looking at psychodynamic theory, a clinician has a template for assessing and treating a client. This paper will discuss depression by examining the DSM-5's explanation for the disorder. This paper will also provide a brief explanation of psychodynamic theory and how this theory, when combined with the DSM-5 diagnosis, can provide a foundation and context for symptoms when assessing individuals who suffer from depression.

Explanation of Depression

In the DSM-5, depression is found in the chapter on Depressive Disorders (American Psychiatric Association, 2013). In order to be diagnosed with depression, five or more symptoms from criteria in category A must be present during two consecutive weeks in which a change from previous functioning has occurred (American

Psychiatric Association, 2013). One of the symptoms has to be either depressed mood or loss of interest or pleasure and it must be present nearly every day for most of the day (American Psychiatric Association, 2013). The following are some of the symptoms:

> The individual may feel a loss of interest and pleasure in almost all activities. Individual may also suffer from changes in appetite, weight, sleep, and psychomotor activity, decreased energy, feelings of worthlessness, guilt, trouble thinking, trouble concentrating, trouble making decisions, irritability, or suicidal thoughts and in some cases, sadness may be denied (American Psychiatric Association, 2013, p. 155).

Symptoms also must cause severe distress or impairment in social, occupational, or other important areas of functioning, such as relationships with others (American Psychiatric Association, 2013). In other words, due to the depression, an individual may not be able to get through the day because of the overwhelming effect of the depressive symptoms. Symptoms also cannot be attributed to other physiological effects of a substance or to other medical conditions (American Psychiatric Association, 2013). The diagnosis of depression cannot be given if there is another explanation for the symptoms such as drug/substance use or the existence of another medical condition.

The DSM notes that when a person suffers a great loss such as a death, financial ruin, loss from a natural disaster, or a serious medical situation, they may show symptoms of depression. The clinician is advised to be careful and to use judgment based on the client's history so as to decide whether the diagnosis of depression is correctly given (American Psychiatric Association, 2013). Every clinician should be careful when assessing an individual regardless of present situations that may cause temporary symptoms of depression. Fatigue and sleep disturbance are often present, which can contribute to an individual not being at a functioning level (American Psychiatric Association, 2013).

In terms of prevalence, 18-29-year-olds have a higher risk and individuals aged 60 or older have three-times higher rate of getting a depressive disorder, although in the United States, there is a peak with individuals in their 20's (American Psychiatric Association, 2013). Sadly, females have a one and a half to three times higher rate than men (American Psychiatric Association, 2013). In terms of recovering from depression, some individuals rarely, if ever, go through remission (2 or more months with no symptoms) (American Psychiatric Association, 2013). Recovery usually begins three months after the initial onset for 2 in 5 people or within one year for 4 in 5 people (American Psychiatric Association, 2013). It seems if a person is suffering from depression, there is a good chance of them recovering within a year but most likely not before three months of being symptomatic. Some risk and prognostic factors for depression include: temperament, environment, genetics and physiology, and course modifiers (American Psychiatric Association, 2013). Course modifiers refer to an individual who suffers from depression, but the depression itself is in the background of another non-mood disorder. Depression is a serious disorder with a high potential for debilitation. Viewing this disorder through several lenses is the best way to assess and devise treatment.

Explanation of Psychodynamic Theory

The origins of psychodynamic theory are from Freud's original theory of psychoanalysis, in which it is thought that early life experiences help shape and mold the personality. Psychodynamic theory recognizes the strength in the powerful dynamics in the "parent-child relationships and in the exploration of the lifeworlds of children" and how those early childhood experiences are central in the patterning of an individual's emotions and, therefore, central to the problems in living throughout life (Hutchinson & Charlesworth, 2003). What an individual experiences early in life sets up a template of how the individual thinks the world and the people in it will interact with him/her.

Psychodynamic theory is concerned with how internal processes such as needs, drives, and emotions motivate human behavior (Hutchinson & Charlesworth, 2003). Essentially this theory looks at all the influences in a person's life to see how this person acts and interacts with the surrounding world and provides insight to the individual's psychological world. Psychodynamic theory has changed over the years moving from a classical psychoanalytic emphasis on innate drives and unconscious processes toward a greater focus on the adaptive capacity of people and how they interact with the environment (Hutchinson & Charlesworth, 2003).

As clinicians who look at person-in-environment, social workers look at all the factors that motivate individuals, so this shift from innate drives to interactions with the environment incorporates all the aspects social workers consider when assessing and treating individuals. The shift makes this theory more holistic. Looking specifically at attachment theory (which falls under psychodynamic theories), the explanation of the role of attachment across the lifespan can be used when assessing the underlying causes of the expression of particular depressive symptoms (Coady & Lehmann, 2008).

Attachment theory founder John Bowlby believed "children's mental health was severely compromised by early separation," which shows the importance of early interactions and the life-long impact they have (Coady & Lehmann, 2008). Clinicians in the field of social work know that an adult's mental health is also compromised throughout the lifespan by early and/or unplanned separations in vital relationships, which will give the clinician insight to the thought patterns and behaviors of individuals. This allows clinicians to see the influence of attachment in both healthy development (the absence of symptoms) and emotional disturbances (presence of symptoms) and then work with the client in developing an effective treatment plan (Coady & Lehmann, 2008).

Combining Depression with Psychodynamic Theory

The DSM-5 is essentially a tool and should be viewed as such when being used by a clinician. The diagnostic tool should be considered a foundation when looking at depression. The clinician assesses the individual and is given a template to see what the symptom clusters mean (i.e., a diagnosis) and with this diagnosis, the clinician then pairs a theory that best applies to the individual. There is now a framework to build off of in which the DSM-5 gives the symptoms and the theory provides a context in which those symptoms are applied to a behavior and/or past trauma. The DSM-5 gives the guidelines, and the psychodynamic theory provides the context in which those symptoms have a meaning that correlates with a behavior. Based on this logic, depression can be seen as needs that were not met early on in life and then this gap is reinforced throughout the individual's life. Depression is considered the most widespread emotional disorder and the leading cause of disability in the world, and while the DSM-5 provides information about what the symptoms are, psychodynamic theory allows the clinician to see what the symptoms look like while in motion (Berzoff, Flanagan, & Hertz, 2011).

Psychodynamic theory suggests that external factors and situations along with skewed perceptions of the environment contribute to the onset of depression. Using psychodynamic theory as the logic, those skewed perceptions can be attributed to possible negative interactions early in life which then created a negative template for later interactions. Using the psychodynamic theory as a framework, individuals can be seen as becoming overwhelmed by internal or external demands and may use ego defense mechanisms to avoid becoming overwhelmed (Hutchinson & Charlesworth, 2003). In other words, according to this theory, when individuals become depressed it is in response to not being able to cope with something, and the response is to completely shut down in order to protect themselves. Self-psychologists think that the symptoms of depression serve to protect a self-esteem that can no longer function and meet its goals (Schwartz & Schwartz, 1993).

It was once thought, "depression is hostility turned inward" or "refers to a situation where a person invests love in another person, and then discovers the relationship terminated" (Hutchinson & Charlesworth, 2003; Schwartz & Schwartz, 1993). Early on it was deemed that depression was an expression of anger and rejection. Ego psychology views losses as injuries more to the individual's view of himself (his ego), therefore any loss could be seen as meaning something personally against the individual (Schwartz & Schwartz, 1993). Going by this frame of thought, it is easy to see how loss and/or rejection can play a role in depression among individuals. If an individual finds meaning in something and then that something is gone, that person might no longer have meaning in his/her life. A central characteristic of depression is a feeling that life is meaningless. Goals that usually drive and motivate a person are no longer there as a focus for ideals and ambitions (Schwartz & Schwartz, 1993). Depression drains the person of what has meaning, and therefore goals become irrelevant.

One limitation of using psychodynamic theory with the DSM-5 is that the DSM-5 excludes certain characteristics of rumination, almost dismissing it by attributing it as a side note about other situations that can cause sadness (but not depression). Research has found that time spent ruminating increases activity in the brain's fear system, which increases avoidant behavior making it less likely a person will derive pleasure from life or emerge from depression (Hellerstein, 2011). The act of rumination interferes with problem solving and makes a person's mood worse, not better (Hellerstein, 2011).

Although the DSM-5 essentially dismisses the act of rumination, the clinician can use this as an assessment tool to see why the individual is not able to get past these particular thoughts. The act of dwelling on something is worth looking into from the point of view of the clinician, and applying psychodynamic theory would encourage the clinician to look at past interactions with family, caregivers, relationships, and the world and use them to explain ruminating thoughts.

A second limitation with using this theory is the fact it "does not have practice principles at the community, organization, or social institution level" (Hutchinson & Charlesworth, 2003). Clinicians look for ways to integrate a positive interaction between the individual and the community and this theory is lacking in that area. While there is strength in looking at the individual, there is no compass in psychodynamic theory for how to apply practice between the client and his/her community for a goodness of fit.

An essential part of the job for social workers is to find ways for clients to overcome difficulties and learn to positively interact with their communities. While psychodynamic theory provides social workers with a framework to assess and treat the individual, it does not have a way to implement treatment so it applies on a macro level. There is a context to treat the client on an individual level, but not on a community level. Also, psychodynamic theory allows the clinician to view depression from the psychological and emotional aspects, but does not consider any biological aspect of depression, such as a possible chemical imbalance. Symptoms are attributed to psychological factors based on interactions with others and the surrounding world. Research has shown that there are "abnormalities in specific areas of the brain, specifically, the memory center, the conflict-resolution area, and the area of the brain involved with planning and executing activities" that are affected when an individual has depression (Hellerstein, 2011). At this point it would be interesting to ponder if there are any risk factors that might contribute to these biological findings which social workers could address in terms of treatment.

Discussion

Although there is significant debate over the usefulness of the DSM-5, it provides a foundation to apply context for a theory. A theory alone cannot provide all the necessary information when assessing and treating a client. The DSM-5 provides the foundation while the theory provides the context in which the individual's symptoms are assessed, allowing the clinician to have a starting point for treatment. While the DSM-5 can be viewed as a cold tool of medicine, psychodynamic theory is the spirit and warmth of humanity. With all modes of assessment and treatments, there are always positives and negatives, and psychodynamic theory will not always be the best theory to use as a template in treating individuals. In my clinician opinion, psychodynamic theory allows the clinician to look at what drives and motivates the clients and how behaviors and thought processes came to be.

Psychodynamic theory and the DSM-5 compliment each other and allow two different perspectives to merge together in order to provide a more comprehensive look into the inner workings of individuals who are in need of assessment and treatment. As clinicians we are trained to assess clients and to meet them where they are in terms of treatments, but a starting point and a point of view is also essential. The goal of the clinician is to meet the basic needs of individuals and teach healthy behaviors. Clinicians have a healing effect by addressing the roots of several symptom causations through the therapeutic relationship. The point of view is the theory, which allows the clinician to have a starting point to begin assessment in terms of the direction and approach. Although the DSM-5 is a medical diagnostic tool, it can prove to be quite valuable when used in conjunction with a social work theory. Using the DSM-5 can be seen as a means to the end result. The first and foremost duty is to provide a safe environment in which the clinician has all the available tools to assess the situation, and sometimes the assessment requires many different perspectives in order to get an accurate picture of what is truly going on with the client. Using tools such as the DSM-5 and pairing it with a theory of social work provides an accurate picture of what is happening in the present (presence of symptoms) and how the past has affected the individual (psychodynamic theory) and how to design an effective treatment based off that information in the best interest of the individual.

References

American Psychiatric Association. (2013). *Diagnostic and Statistical Manual of Mental Disorders* (5th Edition ed.). Arlington, Virginia: American Psychiatric Association.

Berzoff, J., Flanagan, L. M., & Hertz, P. (Eds.). (2011). *Inside Out and Outside In* (3rd edition ed.). Eestover Road, Plymouth, United Kingdom: Rowman & Littlefield Publishers, Inc.

Coady, N., & Lehmann, P. (2008). *Theoretical Perspectives for Direct Socila Work Practice: A generalist-eclectic approach.* New York, New York: Spring Publishing Company.

Hellerstein, D. J. (2011, July 14). *Depression and Anxiety Disorders Damage Your Brain, Especially When Untreated.* Retrieved October 20, 2014, from Psychology Today: www.psychologytoday.com/blog/heal-your-brain/201107/depression-and-anxiety-disorders-damage-your-brain-especially-when-untre

Hutchinson, E. D., & Charlesworth, L. W. (2003). Theoretical Perspectives on Human Behavior. In *Dimensions of human behavior: Person and environment* (pp. 46-88).

Ozdemir, Y., Kuzucu, Y., & Koruklu, N. (2013). Social Problem Solving and Aggression: The Roles of Depression. *Australian Journal of Guidance and Counseling, 23* (1), 72-81.

Schwartz, A., & Schwartz, R. (1993). *Depression Theories and Treatments: Psychological, biological, and social perspectives.* New York: Columbia Unversity Press.

An Emotional Depression

Aleks Kico

Introduction

When we think of prolonged depression we automatically begin to look for biochemical and genetic components to attribute the cause of the disorder. In fact, nearly all literature, research, medical studies, and professionals tend to focus particularly on the biology and the psychology of the individual. It is very rare that we look at the significance that our emotions play in our physical make up with daily living. We commonly attribute negative emotions as a result of disorders, however, the focus of alternative healing methods demonstrates the opposite; they illustrate how an imbalance with emotions ultimately cause disorders.

Recent studies indicate that our emotions are incredibly powerful and dictate every single part of our daily life, whether we are aware of it or not. Michael Brown's theory of the seven-year emotional imprint may further demonstrate why depression and other forms of disorders are often times mistreated. In his theory, Michael Brown discusses the significance that our emotions play in our bodies, particularly negative emotions such as fear, anger, and grief, which he demonstrates having a causal point from the first seven years of our existence (2010). He further describes that the notion of present moment fear, anger, or grief, or any triggered emotion derives from unresolved childhood memories.

Ultimately, the incorporation of multiple studies examining the heart and emotions will produce the hypothesis that the manifestation of depression is a deliberately chosen method of communicating our unresolved emotional traumas. The so-called "disorder" is deliberately chosen by our internal system for the purpose of bringing resolution and balance. The goal of this work is to invite us to turn in to the emotional imprint to receive an organic form of resolution instead of coping strategies that only postpone the suffering or cause it to reappear in other forms.

Depression

Depression is devastating and wildly common among us. Currently WHO says that depression is the number four cause of disability in the U.S. and by 2025 it will be number 2, which further demonstrates that our current practice of treatment is not showing any promising signs of efficiency. When considering depression, the cognitive side of depression is learned helplessness. This work will further demonstrate through the use of the emotional imprinting theory how we ultimately choose to develop a disorder as a method of communication.

In Konigsberg's study of grief, he describes how we need to rethink how we view depression (2011). He goes on to describe that depression is nature's way of shutting down the nervous system in order to protect itself from a situation, which appears too much to bear. His findings further help us to reconsider that perhaps depression is not a disorder but a natural form of coping. We often times consider Western medicine as the front runner of healing; however, what we must keep in mind is that the body's natural state is to heal it self, and that it does not require the sort of treatment we provide for it. For example, if you were to break a bone, the doctor can reset that bone for you, the doctor can give you medication to minimize the pain, however, the body is ultimately able to heal the bone itself, not the medication.

A study concluded that it takes an average of 15 minutes of conversation with a depressed person for people to feel an increase of anxiety and depression as a result of the interaction, and ultimately develop hostility toward that person. The people in this study indicated that as a result of the initial interaction, they were less likely to interact in the future with the same depressed person and were more likely to express their negative reactions as a result of that interaction. The depressed people in this study reported that they felt the sense of rejection and expressed a similar disconnect with the counter partner. Since depression and other forms of mood and cognitive disorders are often times intertwined with a stigma, our diagnosis of disorders manual ultimately cripples the individual in his or her immediate environment.

A recent meta analytic review of 148 studies (representing more than 300,000 individuals) concluded that the influence of social impact is comparable on a mortality risk scale magnitude with that of other commonly recognized risk factors, such as smoking, excessive alcohol consumption, obesity, and lack of physical activity (Holt-Lunstad, Smith, & Layton, 2010). So the lack of social connection was linked to complications in health (particularly cardiovascular). If Western medicine and our culture attribute a disorder label for depression, are we also causing social dissonance?

Seven-Year Emotional Imprinting Theory

Michael Brown describes that our emotions were and continue to be our primary from of communication with the physical world. In his book: *The Presence Process,* Michael describes that in the first seven years of our experience, our only form of communication with our caregiver and the environment was through our emotions. The way we began to communicate with our caregiver began at about two months. At two months in the womb, we received communication with our mother constantly; the emotions our mother was experiencing were expressed throughout her body and passed onto the womb. This statement will be explained in greater detail in the following paragraph, and how it relates to our emotional imprint.

Since the first organ we develop was our heart, that was our very first method of communication with our mother. For example, if the mother were to experience a sudden rush of fear due to a real or perceived environmental circumstance, the body would shift into sympathetic mode in order to preserve the vitals (heart), and thus communicate that response onto the child. Through a variety of emotions experienced during pregnancy, the child begins to develop patterns that subsequently shift heart rate regulation during specific environmental stimuli. This is where we inherited all our particular emotions, including instinct, intuition, urges, likes, dislikes, etc. After birth and up to about age seven we as children expressed the whole range of emotions in order to communicate with our caregiver. At age seven when we were sent off to school, our emotional expressions were culturally encouraged to be controlled (shamed), hidden, or at the very best sedated with medications. We as children were reinforced to respond in a specific way.

Studies of behaviorism such as that of John B. Watson and psychologists alike are further demonstrations of how we parent our children through a reward and punishment styles, which ultimately creates more harm than good. Brown describes that during the next seven years (7-14) we developed our mental body. This is where we learned to verbalize, write, and began to conceptualize social patterns in our immediate environment. The following seven years (14-21), we developed our physical body. This is where we matured physically; we went through puberty; we form relationships, and that is why our 21st birthday indicates coming into full growth.

There are many dimensions to this theory, however, the difficult part Brown attempts to explain is that for the rest of our lives we only develop physically, and mentally, in other words, we are physically mature, we are mentally mature, however, our emotional body remains at age seven, no matter how old our physical body may be. Thus, any form of fear, anger, or grief that we experience in our day-to-day life can trigger all those unresolved subconscious memories to reappear. Any seemingly random discomfort we encounter subconsciously regulates our heart to the pattern we established in the womb.

In fact, Michael Brown demonstrates through guided meditations, how present moment negative reactive feelings are in fact founded in our earliest infancy years, but it is our deliberate attempts to cover up these emotions, which give the illusion of a disorder appearing in the present moment. This theory describes that all physical forms of manifestations (disorder, illnesses, diseases, infections, etc.) have a causal point in the emotional body, and not the biology and psychology of it. His theory further invites us to reconsider methods, which will help us to give rebirth to our emotional child and allow it to mature.

Another interpretation we can add to this dimension to provide an additional perspective is through Einstein's theory of relativity (E=MC2). In this theory it is described that we can consider our self and everything in our physical environment as energy in motion. He demonstrated that atoms vibrate at such a high frequency that our body and everything around us appear to be as a solid object when in fact our body and everything around us, is not; we are simply energy in motion. The reason why this information becomes important to us is because another word for **e**-motion is energy in motion= emotion.

If we can begin to consider our bodies as a natural healing agent of all illness and disorders, we can also begin to consider that perhaps the reason why we are in depression is that the body is attempting to heal itself and restore a balance to our energy (emotions). The challenge becomes in learning to shape our practice to accommodate the full expression of depression instead of controlling it with artificial methods to accommodate cultural norms. Simply put, no one has ever died of depression, however, people die because of shame every day. If we shame our emotions, eventually they all lead to a premature death.

How Emotions Communicate

In her newsletter, *Health Wisdom For Women*, Dr. Christine Northrup writes that our brain and immune system communicate in two ways: by means of hormones that our brain regulates and through protein molecules called neuropeptides and receptors, which send messages back and forth. These same molecules are not only in your brain, but also expand to every tissue, organ and cell of your entire body. Since the network expands to every single cell of our body, it means that every thought you think and emotion you feel is communicated to every cell, organ, and tissue in your body. A thermal imaging conducted of people experiencing the most common forms of emotions was taken to demonstrate how our emotions would look like in our body.

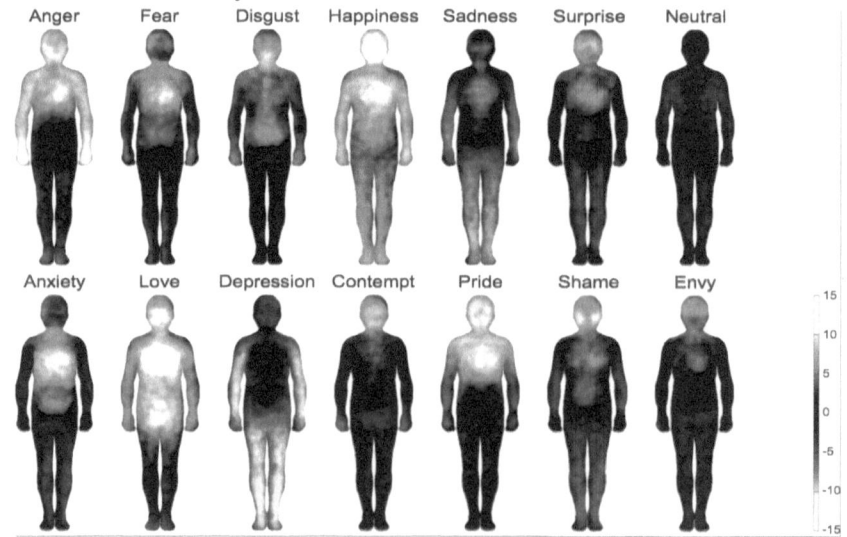

This is where we can begin to develop a clearer sense of how our mother's heart rate response to the environment imprints the child in

the womb. Ultimately, science cannot measure our energy, what science can measure are the symptoms of blocked energy. This is how illness, disorders, diseases, infections, disabilities, and afflictions are formed; its causal point lies in our emotional triggers, and if left unattended to, like a child would, they will cry out until it gets the attention it requires. Since our emotional seven-year-old child has been stunted of growth, the only way to get our attention is through physical manifestations, such as depression.

The DSM-V has provided a chart of symptoms, which professional staff can check to determine the severity of the depression and the treatment route; however, this is where we can begin to implement our knowledge of emotions and energy to ask questions such as: what is my energy attempting to accomplish by asking my body to sleep all day, or not sleep very much? What is my energy attempting to accomplish by decreasing my appetite or eating more than usual? What is my positive intention for having no energy? What is my positive intention for not being able to maintain concentration? What is my positive intention for losing interest in all my activities? What is my positive intention for feeling disconnected and numb? What is my positive intention for feeling sad, empty, and hopeless? What is my positive intention for feeling worthless and guilty? What is my positive intention for wanting to die? What is my pure positive intention for feeling…….? Since our emotional body is dictated by the seven year old in us, it will use our body in order to communicate. Again we must constantly keep in mind that the body's natural state is to heal itself not to remain ill, so at every juncture of the disorder, the body is ultimately healing and that we must be able to recognize these symptoms as a deliberate intention produced by our body to accomplish homeostasis and stability.

Diagnosis and treatment

Depression is commonly recognized as a biochemical disorder with a genetic component. People who are chronically depressed or live with bipolar disorder are often told that they have a chemical imbalance, which is commonly recognized as the cause of the disorder. Treatments usually consists of the use of various forms of medication that help to "knock the edge off" in combination with cognitive behavioral therapy. And in the end, the professional is expected to help the client develop coping strategies, and the client would have learned how to manipulate their emotions and environment to interrupt or eliminate the depression from reappearing.

The problem with this method of practice however is that it creates a further dissonance with the individual and the emotional body by implementing strategies that eliminate the undesired emotional triggers and reactions. What ultimately gets accomplished with current methods of treating depression is that the client will experience a decline in symptoms of depression but that emotional signature that is carried will manifest in another physical form to deliver that same charge. The use of medication essentially produces this question: If we make our barometer for success and happiness the idea of escaping certain emotions, or controlling certain feelings, then what constitutes resolution?

The use of medication and coping strategies certainly does not accomplish this, because we don't learn to resolve the issue, we only learn to manage it. That is why we commonly say that coping with a trauma is a "daily battle"; we don't necessarily receive resolution, we receive relief, which makes the impression that we are healing, but we're really not. From the many interviews that I completed with residents during my internship experience, I discovered quite an interesting pattern developing. The conclusion that I came to was that: of the residents who reported feeling depressed in the biopsychosocial interview, nearly all of them were placed on anti depressants or anti anxiety medication at some point, unless the family refused the practice to be carried out. And of those placed on medication, all reported feeling less depressed in future interviews; however, over the span of a few months, they also developed additional physical complications that they did not come in with, for example, they gained an excess amount of weight, reported feeling more dazed throughout the day, exhibited greater social dissociation, and displayed a lower score on the cognition portion of the interview. In conclusion, even though these residents expressed feeling less depressed with their circumstances, they ultimately developed many other physical conditions along the way, many of which were being treated through additional medication.

Therapists, social workers, and professional staff trained for counseling are normally taught to look for strengths (we call this the strengths perspective) as a means of coping with disorders. Strengths include enjoyable activities, friends, family, skills, etc. What this accomplishes on a superficial level is that we are basically teaching the client that when they are overwhelmed with an emotion, they should do an "act", which produces a desired outcome (relief). However well planned out our coping strategies may be, they will never bring the body authentic resolution and chemical balance; our causal point for depression will perpetually linger on until we are no longer able to escape it. Because we are basically helping the client feel better in the moment, we are ultimately postponing that emotional signature to reappear at a later time, and in a different manner.

Benefits of Tears

Although I have not outlined any concrete steps towards treating depression, nor will I, to avoid this taking the image of a "doing", however, let's consider one of the most commonly shamed natural forms of relief that our body has provided us with. The shedding of tears is more important than we realize; according to doctor's reports, tears contain encephalin, which can decrease physical and emotional pain when it becomes too much to bear. Tears release ACTH, which generates a calming effect. Because we commonly associate tears with pain, we subsequently don't like to see people cry. This notion of emotional relief becomes compounded by shame when we consider gender. Understandably, we don't want to see our loved one depressed and sad, or angry and fearful. But it's not about preventing our loved one from experiencing any undesired emotion; it's about allowing them to organically process that emotion. If that means the expression of tears on a daily basis, we are experiencing a natural form of relief and healing. We as a culture have the formed belief that when we cry, something has gone wrong and that we must change something in order to stop. Unfortunately, we have been indoctrinated to feel happy 24/7, but we do not realize that by the expression of tears, we are receiving organic relief, which is what our body has deliberately provided us with. We must remember that when we cry, it is that seven-year old emotional child receiving attention, and if we shame the tears, we are ultimately shaming that child and restricting it of growth and maturity.

Conclusion

In conclusion, I hope to invite you to reconsider the disorders from an emotional perspective and how we ultimately chose to experience the manifestation as a form of communicating and resolving the emotional charge we carry. Based on my own independent studies and my own professional experience of working with clients who are depressed, it is my personal opinion that one of the most effective methods we can implement in our practice is to facilitate a safe environment for the client to be depressed for as long as they need to without incorporating any methods to bring about relief. As Brown stated in his book, ultimately "the journey of emotional processing is not about feeling better, it's about getting better at feeling". What we essentially offer the client is the opportunity to parent their own emotional child; which ultimately teaches them to validate their emotions instead of escaping or controlling them. To classifying depression as a disorder ultimately creates the amplifying belief that the human body is coincidental and is therefore incapable of tending to its own needs. Some of our greatest assets are our emotions, so how can we recognize and utilize them to our benefit?

References

Ask Dr. Sheryl: Tears--A Natural Stress Release. (n.d.). Retrieved December 6, 2014, from http://drsherylwagner.blogspot.com/2011/03/tears-natural-stress-release.html

Brown, M. (2010). *The presence process: A journey into present moment awareness* (Rev. ed.). Vancouver: Namaste Pub.

Depressive Disorders. (2013). In *Diagnostic and statistical manual of mental disorders: DSM-5*. (5th ed.). Washington, D.C.: American Psychiatric Association.

Holt-Lunstad, J., Uchino, B. N., Smith, T. W., & Hicks, A. (2007). On the importance of relationship quality: The impact of ambivalence in friendships on cardiovascular functioning. *Annals of Behavioral Medicine, 33*, 278–290.

Konigsberg, R. (2011). *The truth about grief: The myth of its five stages and the new science of loss*. New York: Simon & Schuster.

Northrup, C. (n.d.). Women - Understanding the Body-Mind. Retrieved December 6, 2014, from http://www.enotalone.com/health/4666.html

Photo credits: Just Working with My Spirit. (n.d.). Retrieved December 6, 2014, from http://gloriaromlewski.com/tag/positively-oriented/

An Attachment Theory Perspective on Depressive Disorders

Anne Weldon

The Diagnostic and Statistical Manual of Mental Disorders, Fifth Edition (DSM-V) is the 2013 revision of the long-standing tool utilized by the mental health community to classify and diagnose psychiatric disorders. While the latest version has come under fire as lacking empirical evidence, it is still widely regarded as the standard by which mental health is measured in this country. Since reimbursement to health care providers is largely determined by diagnostic code, the DSM-V is likely not going anywhere. That said, there is much debate as to whether the criteria set forth in the DSM-V are an adequate measure of individual mental health. In this paper, I will examine in detail the section on depressive disorders, and specifically the diagnosis of major depression. I will then argue that the criteria set forth in the DSM-V are not sufficient to determine whether a person is truly suffering from depression without further considering the individual's experience of the world around him or her in a more holistic manner.

The World Health Organization described depression as:

> a common mental disorder, characterized by sadness, loss of interest or pleasure, feelings of guilt or low self-worth, disturbed sleep or appetite, feelings of tiredness, and poor concentration. Depression can be long-lasting or recurrent, substantially impairing an individual's ability to function at work or school or cope with daily life. At its most severe, depression can lead to suicide. (www.who.int/topics/depression/en/, 2012)

There are eight sub-categories of depressive disorders described by DSM-V, all of which share one common feature: "sad, empty, or irritable mood, accompanied by somatic and cognitive changes that significantly affect the individual's capacity to function" (American Psychological Association, 2013). Diagnosis for these disorders is based on a patient meeting a specific number of criteria for a certain amount of time.

Diagnostic Criteria

To be diagnosed with Major Depressive Disorder, the DSM-V requires that at least five of the following symptoms are present during the same two-week period with at least one being depressed mood or loss of pleasure:

1. Depressed mood most of the day, nearly every day, as indicated by either subjective report or observation made by others
2. Markedly diminished interest or pleasure in all, or almost all, activities most of the day, nearly every day (as indicated by either subjective account or observation).
3. Significant weight loss when not dieting or weight gain or decrease or increase in appetite nearly every day.
4. Insomnia or hypersomnia nearly every day.
5. Psychomotor agitation or retardation nearly every day (observable by others, not merely subjective feelings of restlessness or being slowed down).
6. Fatigue or loss of energy nearly every day.
7. Feelings of worthlessness or excessive or inappropriate guilt (which may be delusional) nearly every day (not merely self-reproach or guilt about being sick).
8. Diminished ability to think or concentrate, or indecisiveness, nearly every day (either by subjective account or as observed by others).

9. Recurrent thoughts of death (not just fear of dying), recurrent suicidal ideation without a specific plan, or a suicide attempt or a specific plan for committing suicide (APA, 2013).

Prevalence

The DSM-V estimates, "twelve-month prevalence of major depressive disorder in the United States is approximately 7%" (APA, 2013). It asserts that prevalence is three times higher in 18-29 year-olds than in individuals 60 or older and that females are 1.5 to 3 times more likely to develop depression than males beginning in early adolescence (APA, 2013).

Depression diagnosis is largely dependent on self-reported symptoms, often by completing a survey, such as the Patient Health Questionnaire (PHQ-9), a list of nine questions that requires an individual to rate how frequently in a two-week period they have experienced the symptoms listed. This is problematic, in that most of the aforementioned symptoms (with the exception, perhaps, of recurrent thoughts of death) are experienced by many, if not most individuals at some point in time. Depending on the circumstances at the time of evaluation, basing a diagnosis solely on the presence of these criteria could result in over-diagnosis, which in turn could lead to over-medicating. Conversely, the diagnosis of depression carries a social stigma, which may cause an individual to minimize the frequency or severity when reporting his or her symptoms.

In *Attachment and Loss*: Vol. 3, John Bowlby (1980), the founder of attachment theory, wrote:

> Sadness is a normal and healthy response to any misfortune. Most, if not all, more intense episodes of sadness are elicited by the loss, or expected loss, either of a loved person or else of familiar and loved places or of social roles. A sad person knows who (or what) he has lost and yearns for his (or its) return. Furthermore, he is

likely to turn for help and comfort to some trusted companion and somewhere in his mind to believe that with time and assistance he will be able to re-establish himself, if only in some small measure. Despite great sadness hope may still be present. Should a sad person find no one helpful to whom he can turn, his hope will surely diminish, but it does not necessarily disappear. (p.245)

Because the lines between normal feelings of sadness and loss and a diagnosis of depression are blurred at best, it is even more critical that the larger picture be examined. While DSM-V does allow for exclusion due to other potential co-existing conditions, it fails to consider the person outside of the realm of the diagnosis of depression. This is problematic in that the diagnosis of depression can be very stigmatizing. The individual is depressed, therefore, his or her behavior and experiences are viewed through the lens of that diagnosis, rather than considering the diagnosis as a result of the individual's global experience of the world around him or her. It is perhaps more prudent to consider this global experience when an individual presents with symptoms indicative of a depressive disorder, rather than labeling with a diagnosis. I would argue that employing the person-in-environment principle is particularly important when considering this diagnosis. Specifically, looking at depression through an attachment perspective: how might one's attachment pattern affect one's response to the environment?

Origins of Attachment Theory

Bowlby developed attachment theory from the belief that early child-parent interactions influence the course of development. He defined attachment as "a strong emotional tie to a person (or persons) that promotes the young child's sense of security" (Davies, 2011). Attachment theory further developed based on the idea that humans are compelled by relationships rather than internal drives. Infants are born with a predisposition to attach to caregivers, and this is the foundation for how a child develops and organizes his or her relationships going forward (Bettmann, 2006).

In addition to providing a sense of security, attachment also serves to regulate affect and arousal. In a secure attachment, if an infant is over-stimulated or experiencing a heightened state of arousal to the point of distress, he will alert his caregiver, who in turn then helps him to regulate that feeling. Over time these interactions enable the infant to develop the ability to regulate arousal through his own efforts. Conversely, children who have not been helped to develop this ability through a secure attachment, may have trouble controlling their impulses and emotions down the road (Davies, 2011). Or, if parents or caregivers have responded negatively to an infant's distress signs, the child may develop the tendency to inhibit strong feelings in order to maintain the attachment. Thus, they internalize a method of overregulating, minimizing and avoiding strong emotions (Magai as cited in Davies, 2011.) Attachment theorists posit that if an individual's attachment pattern becomes disorganized in early childhood, it becomes the foundation for how he or she will experience and interact with the world around him or her. Bowlby (1980) discussed depression as it relates to a loss and suggested that cessation of interchange between an individual and the external world is what results in depression. As long as that interchange is maintained, the individual may experience frustration, anger, fear, or any number of emotions, but does not experience depression (p.245).

Much research has been conducted on the relationship between insecure attachment patterns and depressive disorders. These studies have confirmed that Bowlby's assertion that insecure attachment patterns, specifically anxious and ambivalent attachment, and depression are linked (West & George, 2002). Because there are several different tools of measurement for adult attachment, it is difficult to draw a decisive conclusion about the role attachment patterns have in depression. That said, it could certainly be useful to consider these when evaluating an individual experiencing depressive symptoms.

Internal Working Models and Agency of Self

Two concepts in attachment theory are particularly useful when looking at depression. Bowlby asserted that over time, a child internalizes a working model (IWM) based on his or her experience of how relationships work and what he or she can expect of others in terms of responsiveness and care (Davies, 2011). IWMs are internalized representations of the "self" and "other" based on a child's interactions with their main caregivers (West & George, 2002). "The dyadic infant-caregiver organization precedes and gives rise to the organization that is the self. The self-organization, in turn, has significance for ongoing adaptation and experience, including later social behavior. …Each personality, whether healthy or disordered, is the product of the history of vital relationships" (Stroufe, 1989, p.71, as cited in Davies.) These models are either confirmed or altered based on what the child experiences in his relationships over time. Attachment theory suggests that IWM's become unconscious filters and organizers of our perceptions of relationships, and that they influence how we behave with others and guide relationships (Bowlby as cited in Davies, 2011). If our IWM guides us in our interactions with the outside world, it stands to reason that it also impacts our reactions to it. Thus, understanding these models can be useful in assessing depressive symptoms.

Sable (2000) claimed that environmental deficits creating insecure attachments in childhood can result in incoherent working models that continue into adulthood. Further, she writes that emotional distress reflects the internalization of adverse experiences, particularly those relating to security and self-reliance. She suggests that symptoms of depression and other mood disorders are responses to disruptions of personal bonds or problematic internal working models rather than just biochemical imbalances (p.56).

In a study of 299 undergraduate students, Wei, M., Shaffer, P., Young, S., & Zakalik, R (2005) examined basic psychological needs satisfaction as a mediator between adult attachment anxiety and attachment avoidance and distress. The results of this study indicated that basic psychological needs satisfaction is more of a mediator for attachment avoidance than it is for attachment anxiety. The researchers attribute this to a function of internal working models of self and others. They suggest that individuals with attachment anxiety tend to have a negative working model of self, leaving them likely to suppress feelings and needs because they feel that they are part of what makes them unlovable (Wei, et al. 2005). It is easy to see, then, how this IWM could manifest in symptoms consistent with a diagnosis of depression. However, individuals with attachment avoidance tend to have a negative working model of others because they have learned that others are unable to fulfill their needs, and they tend to rely on themselves to manage their distress (Wei, et al. 2005.)

This begs the question: if attachment patterns are internalized in infancy and early childhood, can they be changed? Bowlby suggested they can. He felt that depression was the result of the breakdown of the IWM, and that once the belief system is dismantled, it can then be rebuilt (Bowlby, 1980). Agency of self refers to an individual's ability to consciously examine and re-categorize past attachment experiences. West and George (2002) suggest that this ability is critical to mental health and have explored what implications failure of this ability has as it relates to depression. The results of their study found that individuals with preoccupied attachment are often unable to achieve an integration of attachment experience and memory, resulting in a lack of coherence, which in turn contributes to a lost sense of efficacy of the agency of self. They suggest that from this perspective, depression is representative of the frustration of these individuals in their efforts to achieve a coherent representation of attachment (West & George, 2011).

What is DSM-V Missing?

By looking at an individual only in terms of whether or not they meet the criteria as presented in DSM-V, we run the risk of reducing the individual to a diagnosis. DSM-V fails to consider that every individual has a unique set of circumstances that contributes to the way they experience the world around them. The diagnostic criteria don't allow for differentiation based on individual experience. In addition, there is a major stigma associated with depression. Considering that depression is characterized by feelings of sadness, worthlessness and loss of pleasure in everyday activities, being labeled with the diagnosis and subsequent stigma may be more harmful than helpful. Additionally, the diagnosis implies that something is "wrong" with the individual rather than allowing room for the possibility that the depressive symptoms may in fact be a reaction to the external world.

Another potential consequence of the oversimplification of the diagnosis of depression in the DSM-V is that medication might be too readily suggested. By some estimates, 1 in 10 Americans currently takes an antidepressant (Pratt, 2011). While certainly warranted in some cases, I would argue that prescribing an antidepressant is a quick fix to address the "what" but not the "why". Considering that psychotropic medications can potentially have significant adverse effects, they must be monitored appropriately. It is estimated that 50% of people who are treated for depression are treated in primary care settings (Docherty, 1997). While primary care physicians are certainly competent to prescribe anti-depressant medication, they are not mental health specialists, and it would be difficult to a conduct a comprehensive mental health assessment in this setting.

Conclusion

The DSM-V is a good starting point as diagnostic tool. It provides a decent overview of what depression looks like, but in a purely one-dimensional manner. Because of the potential risk of suicide, however, it is critical that depression is diagnosed and treated in a comprehensive manner. Through the lens of attachment theory it is possible to look at the individual in relation to the world around him or her, and to view his or her symptoms in that light, rather than just within the scope of mental illness. If we agree that attachment patterns are ingrained in early childhood and continue to play a major role in emotional development, it is then necessary to consider these patterns when assessing an individual's mental health.

References

American Psychiatric Association. (2013). Diagnostic and statistical manual of mental disorders (5th ed.). Arlington, VA: American Psychiatric Publishing.

Bettmann, J. E. (2006). Using attachment theory to understand the treatment of adult depression. Clinical Social Work Journal. DOI: 10.1007/s10615-005-003-1.

Davies, D. (2011). Child development: A practitioner's guide. New York: Guilford Press.

Docherty, J.P., (1997). Barriers to the diagnosis of depression in primary care. Journal of Clinical Psychiatry 58(1), 5-10.

Gilbert, P. (2013). Attachment theory and compassion focused therapy for depression. In Danquah, A.N., Berry, K. Attachment theory in adult mental health: A guide to clinical practice. Hoboken: Taylor and Francis.

Pratt, L., Brody, D., & Gu, Q. (2011) Antidepressant use in persons aged 12 and over: United States, 2005–2008. Retrieved from http://www.cdc.gov/nchs/data/databriefs/db76.htm

Wei, M., Shaffer, P., Young, S., & Zakalik, R. (2003). Adult attachment, shame, depression, and loneliness: The mediation role of basic psychological needs satisfaction. Journal of Counseling Psychology 52(4) 591-601.

West, M., George, C. (2002). Attachment and dysthymia: The contributions of preoccupied attachment and agency of self to depression in women. Attachment and Human Development 3(4), 278-293.

World Health Organization. (2014). Health Topics: depression. Retrieved from http://www.who.int/topics/depression/en/

Differentiating Anxiety and Fear from Anxiety Disorders within the Undocumented Immigrant Community through Social Constructionist Theory

Sarah Mesick

Anxiety and fear are common feelings which everyone experiences throughout their life. For example, studying for a test, preparing for a new job, or traveling to a new city can cause feelings of fear and anxiety which are developmentally normal. Anxiety and fear can also present in societal situations related to discrimination, oppression and stigmatization. For example, a homosexual male may be fearful to come out to his homophobic father; a black male may be fearful to encounter white police; or an undocumented immigrant may be anxious about living in a new country without proper status.

The Diagnostic and Statistical Manual of Mental Disorders Fifth Edition (DSM-V) looks deeper into these feelings and experiences of fear and anxiety in order to provide clear and cogent criteria to determine if an individual has an anxiety disorder. According to the DSM-V (2013), anxiety disorders "share features of excessive fear and anxiety and related behavioral disturbances" (APA, p. 189). Fear is related to "surges of autonomic arousal necessary for fight or flight, thoughts of immediate danger, and escape behaviors"; while, anxiety is associated with "muscle tension and preparation for future danger and cautious or avoidant behaviors" (APA 2013, p.189). The anxiety disorders that the DSM-V define include separation anxiety disorder, selective mutism, specific phobia, social anxiety disorder, panic disorder, panic attack specifier, agoraphobia, generalized anxiety disorder, substance/medication-induced anxiety disorder, anxiety disorder due to another medical condition, other specified anxiety disorder and unspecified anxiety disorder. By providing clear diagnostic criteria for the aforementioned disorders, the DSM-V attempts to help clinicians differentiate developmentally normal anxiety and fear from an actual anxiety disorder.

The DSM-V (2013) states that "anxiety disorders differ from developmentally normative fear or anxiety by being excessive or persisting beyond developmentally appropriate periods" and by being "developmentally inappropriate" (APA, p. 189). This statement to distinguish normative fear and anxiety from a disorder plays an important role in determining whether someone is diagnosed with an anxiety disorder or not. As a clinician, the primary questions to consider are "what is developmentally normal?", and "what are developmentally appropriate periods?" These questions are centered specifically on the client's psychological stages of development and are further defined through diagnostic descriptions in the DSM-V.

While the DSM-V's statement to differentiate developmentally normative fear or anxiety from an actual anxiety disorder is useful for clinicians, it is limited and fails to acknowledge the role of socially constructed labels and systems that impact the mental health of millions of people living in the United States. This paper will specifically focus on the people who make up the undocumented immigrant community and the anxiety and fear that exists within this community due to social constructed labels and systems. Due to the DSM-V's individual, self-centered focus, clinicians can often misdiagnose their undocumented immigrant clients and place blame on the individual without addressing the socio-cultural factors at hand. By using social constructionist theory to understand the socio-cultural factors that undocumented immigrants are faced with, clinicians can become more aware of the anxiety and fear present in the undocumented community; and, therefore, have a greater ability to differentiate anxiety and fear from actual anxiety disorders.

Mental health practice has increasingly moved in the direction towards understanding individuals as not only in relationship with themselves, but also through their interactions with others and the environments surrounding them. Rather than solely focusing on the individual and using psychoanalytic theories to understand the individual's unconscious "self", practitioners are recognizing the importance of understanding sociological theories to better understand their clients. Social constructionism is a sociological theory representing this movement because it posits that "human beings actively and symbolically construct the world around them" (Jost & Kruglanski, 2002, p.169). For example, social constructionists believe that any person can be defined in innumerable ways and that identity is created through social interaction. The theory continues to explain that deviant action is not inherently deviant, but, rather it is constructed; and, that social problems are socially problematic only if humans claim them to be problematic (Harris, 2006, p. 224).

Social constructionists also address the concept of language and the belief that individuals, couples and families do not inherently create problems. Instead, social communication and the language around human systems (individuals, couples and families) is what creates the problem. According to Rudes and Guterman's (2007) article, *The Value of Social Constructionism for the Counseling Profession: A Reply to Hansen*, "problems are not the result of an objective defect that exists within or between individuals, but, rather, 'the distinction of the system of treatment concern . . . is defined by those who share in the communication that defines a problem' (p. 391). Through a social constructionist lens, the understanding of the individual and self is significantly shifted to place recognition and focus on socially constructed identities, systems and environments in which the individual lives in and the meanings attached to those constructs.

Along with identities, systems and environments being socially constructed, social inequalities themselves receive meaning through human interpretation. According to Harris's (2006) article, *Social Constructionism and Social Inequality,* social inequalities are social constructs that should not be overlooked as "just", or only, social constructs. Harris explains that social constructionism is "not an approach that applies only to (what are thought to be) nonexistent entities or fallacious ideas. If the social construction of reality is the process through which knowledge is created and classifications are applied, then *everything* is socially constructed...serious phenomena (such as rape or poverty) both become recognizable features of the world through processes of social construction" (p. 230). That means that a homosexual male who is fearful to come out to his homophobic father, a black male who is fearful to encounter white police, or an undocumented immigrant who is anxious about being in a new country are all significant, "recognizable features of the world", that should be taken seriously. Given the extensive implications that these social constructs play in our lives, it is essential that there is greater understanding of societal constructs when working with clients.

Social constructionist theory is particularly pertinent for clinicians to consider when working with people who are members of marginalized groups in the United States. This is important because marginalized groups often face higher impacts of inequalities, discrimination, oppression and stigmatization by the dominant culture. An example of a marginalized population that is impacted by socially constructed inequalities is the undocumented immigrant community.

According to the Pew Hispanic Center, there are approximately 11.5 to 12 million immigrants who are undocumented in the United States (Passel, 2006). This undocumented immigrant status signifies that these individuals do not have access to a valid social security number; which, therefore, does not allow them to work legally in the

United States. Due to their inability to obtain a social security number and the fact that some entered the United States without permission, these undocumented immigrants have been labeled as "illegal" by many mainstream sectors of society.

For the majority of undocumented immigrants, there is no way to adjust their immigration status to become "legal" due to the U.S.'s broken immigration system. Due to the lack of options for undocumented immigrants to become "legal", undocumented immigrants now carry the negative, socially constructed label of being "illegal". This label of "illegal" has assisted in associating undocumented immigrants with being "criminals", "dangerous", and "law-breakers". Social constructionists would connect this labeling to the social communication and language that influences the construct of the identity, perception and meanings associated with undocumented immigrants.

Furthermore, the anti-immigrant groups and politicians in the United States have taken advantage of these labels of "illegal", criminals", "dangerous", and "law-breakers" in order to advocate for socially constructed, anti-immigrant policies and programs that target undocumented immigrants. Specifically, by labeling undocumented immigrants as "illegal" that are connected to "criminal behaviors", these anti-immigrant leaders have been successful in expanding the oppressive detention and deportation systems in the United States.

According to the Immigration Policy Center's report (2014), *The Growth of the U.S. Deportation Machine*, the "federal government has been pursuing an enforcement-first approach to immigration control that favors mandatory detention and deportation" (Ewing, p.1). The report further shared that deportations under the Obama administration have reached over two million individuals, and government funding for deportations of undocumented immigrants continues to increase.

The result of the increase in detentions and deportations equates to an increased number of families being torn apart. These individuals are now living with greater fear because family members are at greater risk for deportation. According to the Chronicle of Social Change's (2013) report, How Deportation Impacts the Health of Families, "Deportation policy creates a climate of fear and paralysis in communities. People are afraid to drive, afraid to use parks and exercise outdoors, afraid to use public services like clinics, and afraid to get involved in their communities" (Hellerstein, p. 1).

Furthermore, in most states, undocumented immigrants are not able to apply for a driver's license because they do not have a valid social security number. This inability to obtain a driver's license makes them avoid leaving home, driving their kids to school, and driving to work because they are at risk of being caught by the police which could lead to detention and deportation.

Given the oppressive systems and environments in which undocumented immigrants live in, it is essential that clinicians understand social constructionist theory. As stated earlier, social constructionism recognizes that deviant action is not inherently deviant, rather it is defined as deviant through human interpretation. The reason anti-immigrant advocates and politicians can successfully expand programs and policies that detain and deport immigrants is because of the socially constructed idea that entering the U.S. without permission is a "deviant behavior". Therefore, due to the socially constructed "deviant behavior" by undocumented immigrants, anti-immigrant groups and politicians can justify the socially constructed labels of undocumented immigrants as being "illegal", "criminals", "dangerous", and "law-breakers". These labels are then used to further oppress the undocumented immigrant community and justify the increase of the detention and deportation systems.

Fear and anxiety are two emotions that permeate throughout the undocumented immigrant community due to the aforementioned socially constructed labels and systems. Jose Arreola, an undocumented immigrant from Mexico, explains that the constant fear and uncertainty take a toll on one's mental health. Arreola states, "As human beings our number one emotional need is acceptance. If you're undocumented from the get-go you're not accepted. What does that do to one's sense of agency?" (Lyndersen, 2013). Arreola's statement recognizes the personal impact that the socially constructed label has on an undocumented immigrant. When she spoke in an interview about the anxiety that exists in her life, Mayra Sarabia, an undocumented immigrant from Mexico, stated that it is important for her to recognize that she is as valuable as any other human being (Sarabia, 2014). Through these two statements, it is evident that the sense of agency of an undocumented immigrant is damaged due to society telling them they are not welcome or accepted in the United States.

While undocumented immigrants may also experience feelings of depression or trauma and stressor-related disorders; the feelings of fear and anxiety relate closest to the DSM-V's criteria of anxiety disorders. Several anxiety disorders that undocumented immigrants may identify closely with include separation anxiety disorder, social anxiety disorder and generalized anxiety disorder.

According to the DSM-V (2013), separation anxiety disorder is characterized by "developmentally inappropriate or excessive fear or anxiety concerning separation from those to whom the individual is attached". This disorder closely connects to an undocumented immigrant's fear of being deported and separated from a loved one due to the increase in detention and deportations. Social anxiety disorder is characterized by a "marked fear or anxiety about one or more social situations in which the individual is exposed to possible scrutiny by others"; and, where the "fear or anxiety is out of proportion to the actual threat posed by the social situation". Social anxiety disorder can present itself within the undocumented immigrant community due to fear of being discriminated against in social situations and displaying avoidant behaviors due to fear of getting involved and risk of deportations. Lastly, general anxiety disorder is characterized by "excessive anxiety and worry (apprehensive expectation), occurring more days than not for at least 6 months, about a number of events or activities (such as work or school performance)" (APA, p. 190 - 222). Generalized anxiety disorder connects to the undocumented immigrant's day-to-day fear and anxiety of going to work without a legal status, driving without a driver's license, increased chances of being deported and not having access to equal opportunities as other U.S. citizens or legal permanent residents.

All of these anxiety disorders relate closely to the feelings and experiences of undocumented immigrants which makes it easy for clinicians to mistakenly interpret their feelings and behaviors as an actual anxiety disorder. The DSM-V attempts to differentiate and separate diagnosis based on words and phrases related to "inappropriate", "out of proportion", and "excessive"; however, there is a large grey area for clinicians when working with the undocumented community. The grey area exists because the feelings of fear and anxiety that undocumented immigrants have are appropriate and directly related to the socially constructed, oppressive and negative labels and systems in the United States; however, there is no conversation in the DSM-V related to social constructionism.

As stated earlier, the DSM-V (2013) says that "anxiety disorders differ from developmentally normative fear or anxiety by being excessive or persisting beyond developmentally appropriate periods" and are "developmentally inappropriate" (APA, p.189). This statement is limited because it neglects to address the socio-cultural factors that impact an individual's emotions and behaviors. Furthermore, clinicians can be confused while working with undocumented clients because the fear and anxiety that they face does not fall into the category of being "developmentally normal"; and, at the same time, the feelings are not "inappropriate" and they are a result of actual threats from socially constructed systems and labels that impact their mental health.

The DSM-V (2013) also states that "since individuals with anxiety disorders typically overestimate the danger in situations they fear or avoid, the primary determination of whether the fear or anxiety is excessive or out of proportion is made by the clinician, taking cultural contextual factors into account" (APA, p. 189). Since the clinician has the power to determine the excessiveness and levels of fear and anxiety, it is imperative for the clinicians to be provided with further guidance and context when working with undocumented immigrant clients.

In order to properly diagnose clients, the clinician's take must be able to take into account not only the cultural contextual factors, but the *socially constructed* factors of being labeled "illegal" and living in an environment where deportations and detentions are more common than ever. If a clinician fails to recognize that the undocumented immigrant's feelings of fear and anxiety are appropriate and rooted in actual threats, there could be a misdiagnosis which would place another stigma on the client by labeling them with an anxiety disorder.

In conclusion, the undocumented immigrant community is a significant marginalized population in the United States that confronts high levels of fear and anxiety due to being labeled as "illegals", "law-breakers", "dangerous", and "criminals"; as well as being increasingly targeted for detention and deportation by the federal government. While the feelings of fear and anxiety are highly prevalent within this community and closely related to anxiety disorders such as separation anxiety disorder, social anxiety disorder and generalized anxiety disorder; there is a significant differentiation that must be made because the feelings that undocumented immigrants have are appropriate and related to an actual threat.

The DSM-V fails to discuss the role of socio-cultural factors of the undocumented immigrant community which can lead to misdiagnosis and additional mislabeling of the undocumented immigrant. By incorporating social constructionist theory into the DSM-V's description of anxiety disorders, clinicians will be able to better understand their undocumented immigrant clients and the mental health issues they face due to socially constructed labels and systems.

References

American Psychiatric Association (APA). (2013). Diagnostic and Statistical Manual of Mental Disorders (5th Edition ed.). Arlington, Virginia: American Psychiatric Association.

Ewing, W. (2014). *The growth of the U.S. deportation machine: More immigrants are being "removed" from the United States than ever before*. Immigration Policy Institute.

Jost, J. T., & Kruglanski, A. W. (2002). *The estrangement of social constructionism and experimental social psychology: History of the rift and prospects for reconciliation*. Personality & Social Psychology Review (Lawrence Erlbaum Associates), 6(3), 168-187.

Harris, S. (January 01, 2006). *Social constructionism and social inequality*. Journal of Contemporary Ethnography, 35, 3, 223-235.

Hellerstein, E. (2013). *How deportation impacts the health of families*. Journalism for Social Change Fellows.

Lydersen, K. (2013, April 23). 'That feeling doesn't go away': Mental health and undocumented children [The Reporting on Health Member Blog].

Nwosu, C., Batalova, J., & Auclair, G. (2014). *Frequently requested statistics on immigrants and immigration in the United States.*

Passel, Jeffrey S. 2006. *"The size and characteristics of the unauthorized migrant population in the U.S.: Estimates based on the March 2005 current population survey."* Washington, DC: Pew Hispanic Center.

Rudes, J., & Guterman, J. T. (2007). *The value of social constructionism for the counseling profession: A reply to Hansen.* Journal Of Counseling & Development, 85(4), 387-392.

Sarabia, M. (2014, October 19). *Anxiety within the Undocumented Community* [Video file]. Retrieved from https://www.youtube.com/watch?v=mMi0D9vMgAw

Look Who's Talking

Lizeth Pimentel

Background

In 1877 selective mutism was identified by Adolf Kussmaul as aphasia voluntaria, a childhood disorder characterized by a child's lack of speech in certain environments and with certain people (Cohan, Chavira, and Stein, 2006). As a professional culture, social workers have become more aware of the language applied and the connotation it carries when diagnosing individuals. Professionals have been developing their language and literature in order to be more responsive and inclusive of individuals or populations affected by different disorders. With this in mind one can review the change in the Diagnostic and Statistical Manual of Mental Disorders Fifth Edition (American Psychiatric Association, 2013), which resulted in aphasia voluntaria being referred to as selective mutism, which was in itself a change from elective mutism in the DSM III (Cohan et al., 2006). Selective mutism's initial change in language implied that oppositional behavior--or not speaking--was perceived as a voluntary and purposeful action on behalf of the individual. By altering the language in order to include the descriptor "selective," the diagnosis does not subsequently remove the individual from oppositional behavior. Although "selective" can be interpreted as the individual deciding when, where, and/or whom they would, or would not speak to, the focus on "selective" influences the diagnosis.

Language and its meaning has an impact on the manifestation of how a behavior is acted out by an individual. Taking into account the labeling theory, which states that individuals who are labeled with a disability come to accept and live by the definition of their disability, one must take into account the implication language has as well as an individual's level of cognitive understanding on the behavior expressed (Moore and Li, 2001). The labeling theory goes on further to explain that people who are diagnosed with a disorder may or may not have a thorough understanding of its implications, depending on the language used in the description of the diagnosis and how it is explained and understood by the community. If they do have an understanding of their diagnosis, they may unintentionally enact a self-fulfilling prophecy. If the individual diagnosed does not understand their diagnosis, the community and/or family who do know about the diagnosis can influence the enactment of the behavior that defines selective mutism, by enforcing or not reinforcing certain behaviors. Therefore, the behavior others can impose on the individual may contribute to the development of selective mutism behaviors.

DSM V Diagnosis

The DSM V (American Psychiatric Association, 2013) in the DSM V states the following diagnostic criteria for selective mutism:

A. Consistent failure to speak in specific social situations in which there is an expectation for speaking (e.g., at school) despite speaking in other situations.
B. The disturbance interferes with educational or occupational achievement or with social communication.
C. The duration of the disturbance is at least 1 month (not limited to the first month of school).
D. The failure to speak is not attributable to a lack of knowledge of, or comfort with, the spoken language required in the social situation.
E. The disturbance is not better explained by a communication disorder (e.g., childhood-onset fluency disorder) and does not occur exclusively during the course of autism spectrum disorder, schizophrenia, or another psychotic disorder.

(American Psychiatric Association, 2013, p.195)

As the DSM V inclusively states, children diagnosed with selective mutism do not initiate social interactions when encountering others. The lack of speech interaction is observed in both interactions with children and adults. The DSM V (American Psychiatric Association, 2013) also goes on to describe that a typical example of the social interactions that children with selective mutism exhibit are that they are willing to speak in their home when surrounded by immediate family members, but often do not speak when they are in the presence of second-degree relatives or friends.

Not only are they not verbal with family, children with selective mutism are often confronted with high social anxiety within the school setting. Because "Children with selective mutism refuse to speak at school," it impacts the child's educational and academic accomplishments, as their lack of interaction can lead to a difficulty in assessing reading skills (American Psychiatric Association, 2013, p. 195). It is also reported in the DSM V (American Psychiatric Association, 2013) that children with selective mutism have difficulty in social communication because children with selective mutism typically resort to nonverbal communication (i.e., grunting, pointing, writing) to interact. In addition, it should be noted that not all children with selective mutism shy away from engaging in social interactions. Some do engage in school activities that require social interaction, such as taking part in the nonverbal roles in a school play (American Psychiatric Association, 2013).

Selective mutism has comorbid behaviors. As outlined in the DSM V (American Psychiatric Association, 2013), the comorbidity includes: "excessive shyness, fear of social embarrassment, social isolation and withdrawal, clinging and compulsive traits, negativism, temper tantrums or mild oppositional disorder" (p. 195). There is no direct connection to indicate that children with selective mutism have a language disability, because children with selective mutism "generally have normal language skills, [but] may occasionally have an associated communication disorder" (American Psychiatric Association, 2013, p.195-196). Although there is no specific communication disorder, children with selective mutism are usually diagnosed with another anxiety disorder as well, namely, social anxiety disorder (American Psychiatric Association, 2013). Selective mutism is typically diagnosed in children; the prevalence varies from 0.03% -1% depending on the setting (e.g., clinic, school, and general population), and does not appear to vary among race/ ethnicity or sex (American Psychiatric Association, 2013). Children are usually diagnosed before the age of five years old, but are not brought to clinical attention until they enter school; a setting where there is an increase in social interaction and social performances (American Psychiatric Association, 2013). There have been some reported cases where SM disappears, but there is no definite approach to overcoming selective mutism.

Labeling and Behavioral Theory

Pope and Tarlov (1991) explain that communities view a disability as impairment. Therefore, "impairment" holds the connotation that people diagnosed with a disability have a mental and/or physical limitation. The gap between environmental demands and a person's capabilities becomes magnified due to their cognitive and/or physical limitation, based on their social context (Pope and Tarlov, 1991). In addition, Li and Moore (2001) outline that people categorized with a disability are evaluated by norm violating behaviors. Not only does norm violating behavior serve as a way to stigmatize individuals, but individuals with disabilities are also treated based on their categorical diagnosis rather than on their individual characteristics (Li and Moore, 2001). The categorical description of the diagnosis influences the way individuals with a diagnosis are treated, which then contribute to the individual's behavior. Consequently the individual's behavior might then be a product of the social stigma and reactions that are associated with the diagnosis, or their perception of personal responsibility (Li and Moore, 2001). As viewed through the lens of a behavioral framework, selective mutism is defined as a learned conditioned response (Cohan et al., 2006). The implications of a behavioral conceptualization are perceived as either an escape from anxiety or as a way of gaining attention from others (Cohan et al., 2006).

Theoretical Implications on Selective Mutism

Labeling theory is one of our communities' critical frames of reference for labeling and/or stigmatizing others in the environment. Labeling theory is the core focus through which researchers can understand how individuals are perceived and acted upon by others in their surrounding environment. Individuals with selective mutism restrict to whom they speak and where they speak. Due to individuals with selective mutism placing limitations on the environment and people they speak to, it lead me to conclude that selective mutism is driven by societal norms. These norms are beyond the biological or functional limitations of an individual. Rather, the surrounding environment and people can exacerbate or minimize the diagnosis placed on an individual with selective mutism.

As the labeling theory states, an individual has three choices regarding their diagnosis: deny the existence of the diagnosis, accept the diagnosis, or seek benefits from the diagnosis (Li and Moore, 2001). Yet, it remains difficult for the individual to develop an identity due to the conflicting views from their family, community, and themselves. This difficulty can stem from the fact that, following a diagnosis, the diagnostic criteria of selective mutism is subsequently interpreted by family, teachers, and the rest of the community, which can contribute to an individuals' behavior. In addition, this challenge can also be partly attributed to the individual's understanding of what selective mutism means. Therefore, the association between the diagnosis and behaviors that define selective mutism are accepted by the individual and perpetuated, as they come to build a sense of self.

As stated in the DMS V (2013), selective mutism is typically diagnosed around the age of five when children are enrolled in school. Rogoff (2003) describes school-age children as responsible and teachable, as they begin to take on greater responsibilities according to their cultural values. Sameroff and Haith (1996) discuss that in Western culture, children's availability to more peers and adults, and as a result a greater separation from caregivers, siblings, and home sets new standards for children to cope with their social emotional development. Therefore, when the expectations of children's interactions or communication skills in a specific environment are not met, caregivers look for an explanation, which in turn may or may not lead to a diagnosis. A child's refusal to speak in front of others that are not immediate family may serve as a way to avoid engaging in an aversive situation. Yet, a child's refusal to speak in peer interactions or when asked to speak are seen as norm deviating behavior.

In addition, school-aged children are also expected to ask questions to enhance their learning experience. Not only are five year olds expected to adjust to a new environment where some expectations are new to the child, but they are also being assessed on their reading skills. It is not that children with selective mutism do not speak, but because they are more likely to not be verbal in a school setting, it becomes difficult for educators to assess a child's level of reading. This can greatly impact the child's reading score, which if assessed or not, influences a child's educational track, which can lead to behaviors of a child with selective mutism to either be reinforced or challenged.

Cunningham, McHolm, Boyle, and Patel (2004) report that not speaking may be attributed to a child's transition from home to a different setting that requires altogether more interaction. As individuals develop they build an identity based on a bidirectional effect. Family, peers, teachers, administrators, and the community influence the bidirectional effect (DeHart, Sroufe, and Cooper, 2004). Fischer and Ayoub (1994) write that preschool-aged children's dichotomous thinking is an "affect split," a form of dissociation of thinking about the self and others (as cited Harter, 1999). Preschoolers' inability to think of the self or others in a dichotomous perspective, such as good and bad, nice and mean, or love and hate, can affect the child's internalized self- representation (Harter, 1999). These cognitive limitations affect preschoolers' self-representation because they create a bias, a univalent perspective of the child's overall worth as a person (Harter, 1999). When children are unable to give attributes of opposing valences of the self or others, they create a self -representation based on one experience. Scheff (1974) argues that people who are labeled with a diagnosis internalize the stereotypes. Therefore, the stereotypes contribute to a child's reoccurring behavior, which influenced their self-identity.

This environmental influence and the interplay with a child's cognitive development can cause the child confusion, because he or she learns to associate as being distinctly "good" or "bad." Children's growing cognitive capacity to think in a dichotomous categorization affects their social emotional development as children internalize their self-concept, which then mediates how they participate in social interactions (Harter, 1999). They carry their psychological understanding and inferences about the past into the future, which contributes to the developing identity of not speaking in certain social situations. As children get older and spend more time in an educational setting, their changing self–descriptions shift from dichotomous thinking to having a greater focus on personality (Harter, 1996). Consequently, the behavior their peers reinforce can lead to children continuing to not seek verbal communication because they are still able to interact through nonverbal gestures.

According to Fischer's (1980) skills theory, children five to six organize and structure representations in a process that requires representational mapping, the ability to differentiate, integrate, and coordinate concepts of the self (as cited in Harter, 1996). Children at this age are aware of being observed, "I observe you observing me" (Harter, 1996, p. 230). Therefore, they are aware of their behavior and may continue to perpetuate the behavior that has been reinforced. Within the context of pride and shame the emergence of a child's looking-glass self is apparent in that children internalize the attitudes and opinions of others in their self-representations (Harter, 1996). School-aged children learn how to critique the self, as well as pride and shame, in the absence of others (Harter, 1996). As children become more self-critical, observant, and comfortable in their surroundings they begin to explore and "grow out" of their diagnosis. The child shifts from basic emotions to self-affects that extend beyond happy, sad, mad, and scared- they begin to experience feelings of pride and shame (Harter, 1996).

Given that the average age of diagnosis is between three and eight years old as well as keeping in mind that children's cognitive categorization is dichotomous, we can understand how children diagnosed with selective mutism continue to perpetuate their behavior; which is based on the speech behaviors that are reinforced, what children hear others say about them, and how they eventually "grow out" of it (Baldwin and Cline, 1991). As school-aged children begin to experience a different environment outside of the home, they become more comfortable speaking and interacting with others.

Despite the DSM's organization of information, I would define selective mutism differently. From my clinical professional experience and the labeling and behavior theory framework, I do not view selective mutism as a disorder but rather a learned behavioral response. Many interventions have been used to treat what we currently define selective mutism to be. My understanding is that children need to learn how to interact. Given that children speak at home, but do not speak in other settings or to others who are not immediate family, they do eventually become comfortable enough to speak in such settings. Some children are shyer than others. The expectation from our society that children must speak to their peers in order to fit the "typically developing" child category can mislead people from acknowledging that not all individuals wish to or want to interact in the same manner. I think that as a community we imply and want to label individuals because they stray from the "typical development." But that should not be the case, especially when children that are labeled with selective mutism are excelling in the educational setting and do not have any behavioral problems (Cunningham et al., 2004).

Conclusion

A child's wish to not speak to certain people or in certain situations should be seen as avoiding an uncomfortable situation or a willing behavior instead of the disorder as the DSM V portrays it to be. As various sources of literature state, selective mutism is outgrown. As children develop they speak to people and in environments that they previously did not. Keeping this and the child development literature of Harter (1996), Harter (1999), Rogoff (2003) and Sameroff and Haith (1996) in mind we need to account for the child's cognitive development. Children are learning who they are by building self-representations of themselves from their understanding of what is being said of them by their caregivers and community members. In general people will act how they are perceived to be, or however they choose. At such a young age, children are developing who they are through their interactions with others. When certain behaviors are reinforced, they will continue to act in a certain manner if it means they will still be able to be part of the community. Children with selective mutism, as previously stated, interact with the community and their peers, participate in social events, and do just as well academically as children that are not diagnosed with selective mutism (Cunningham et al., 2004). Overall, we have to be mindful of children's developing understanding of the self and what behaviors are being reinforced. It is natural to see a discrepancy in a child's behavior compared to their age group. But being cognizant of the fact that children ages three to eight years old are typically diagnosed with selective mutism and subsequently are still learning and internalizing what is being said of them into their self-representation, should be the main focuses of helping a child verbalize their interaction with others.

References

American Psychiatric Association. (2013). *Diagnostic and Statistical Manual of Mental Disorders* (5th ed.). Washington, DC.

Baldwin, S., & Cline, T. (1991). Helping children who are selectively mute. *Educational and Child Psychology.*

Cohan, S. L., Chavira, D. A., & Stein, M. B. (2006). Practitioner review: Psychosocial interventions for children with selective mutism: A critical evaluation of the literature from 1990–2005. *Journal of Child Psychology and Psychiatry, 47*(11), 1085-1097.

Cunningham, C. E., McHolm, A., Boyle, M. H., & Patel, S. (2004). Behavioral and emotional adjustment, family functioning, academic performance, and social relationships in children with selective mutism. *Journal of Child Psychology and Psychiatry, 45*(8), 1363-1372.

DeHart, G.B., Sroufe, L.A., & Copper, R.G. (2004). *Child development: Its nature and course* (5th Ed.). New York: McGraw Hill.

Harter, S. (1996). Developmental changes in self –understanding across the 5 to 7 shift. In M.Haith and A. Sameroff (Eds.) *The five to seven year shift: The age of reason and ressibility* (pp. 207- 236). Chicago: University of Chicago Press.

Harter, S. (1999). *The normative development of self –representations during childhood. The construction of the self: A developmental perspective.* (pp. 28-41). New York: Guilford Press.

Moore, D., and Li, L. (2001). Disability and illicit drug use: An application of labeling theory. *Deviant Behavior, 22*(1), 1-21.

Pope, A. M., & Tarlov, A. R. (Eds.). (1991). *Disability in America: Toward a national agenda for prevention.* National Academies Press.

Rogoff, B. (2003). *The cultural nature of human development.* New York: Oxford University.

Sameroff, A., & Haith, M. (1996). Interpreting developmental transitions. In A. Sameroff & M. Haith (Eds.), *The five to seven year shift: The age of reason and responsibility* (pp. 3-15). Chicago: University of Chicago Press.

Scheff, T. J. (1974). *The labeling theory of mental illness.* American Sociological Review, 444-452.

Considering Anorexia Nervosa from a Sociocultural Perspective

Jen Brown

Anorexia nervosa is a debilitating psychological disorder with life-threatening physiological symptoms, which has increasingly grown more common, resulting in a near epidemic seen among predominantly adolescent females. It has become synonymous with self-starvation, and the disorder's history with this particular phenomenon weaves together to create an interesting and complicated dynamic encompassing the presentation of anorexia nervosa today.

The "discovery" of anorexia nervosa is most often attributed to two physicians, Gull and Leseque, who each separately documented the disorder in 1873 (Bemporad, 1996). However, a few years prior, a doctor in France presented cases of what would later be called anorexia nervosa, and self-starvation had already been a documented phenomenon for centuries. In 1859, France's Marce described a group of adolescent females who exhibited an intense conviction that they should not eat, accompanied by an obsession with food and an opposition to treatment (Bemporad, 1996). Years later, in 1868, Gull briefly mentioned a case of "hysterical apepsia" in a paper he delivered, only coining the term anorexia nervosa in 1873. Leseque also wrote of hysterical anorexia in a report published in 1873, but both physicians' descriptions completed the full clinical picture of the disorder originally introduced by Marce and set the framework for how the disorder would be viewed for hundreds of years afterward (Bemporad, 1996).

With its late 1800s' incorporation into the psychiatric and medical fields, anorexia nervosa gradually became a familiar disorder to clinicians, but disordered eating did not yet appear common among females in the early 20th century (Perlick & Silverstein, 1994). However, in the 1920s, the American Medical Association felt it necessary to call an emergency meeting in order to discuss the health risks for female college students suffering from anorexia nervosa, as prevalence rates were then approaching potential "epidemic" rates (Perlick & Silverstein, 1994). Research notes the significant increases that took place in incidence rates of anorexia nervosa during the 20th century, particularly for young women (Keel, 2010). Incidence rates in the United States began to rise during the 1960s and increased significantly throughout the 1970s and 80s (Levine & Smolak, 2010). Some evidence further suggests that rates continued to rise during the early part of the 21st century. Although currently available rates of prevalence and incidence should be looked upon with a cautious eye (Levine & Smolak, 2010), recent reports include lifetime prevalence rates of approximately 0.5% for young females, meaning one in every 200 women will suffer from anorexia nervosa at some point in her life (American Psychiatric Association, 2013; Keel, 2010).

In this paper, I make the argument that anorexia nervosa, as well as the other body image-related eating disorders, are clinically significant disorders and should continue to be recognized as such; however, the societal and cultural contexts in which they thrive are too crucial in defining these disorders to be left out of the conversation.

Anorexia Nervosa: A Closer Look

Diagnostic Criteria

As briefly described above, anorexia nervosa is a serious disorder that, by definition, incorporates three aspects: a persistent restriction of the amount of energy that is consumed by the body, an intense fear of weight gain or a pattern of behaviors that are not conducive to gaining weight, and a distorted perception of one's own weight or shape (American Psychiatric Assocation, 2013). First, the Diagnostic Statistical Manual of Psychological Disorders (DSM) lists the criterion of "restriction of energy intake relative to requirements, leading to significant low body weight in the context of age, sex, developmental trajectory, and physical health" (American Psychiatric Association, 2013, p. 338). It goes on to explain that "significantly low weight" is defined by what is "normal" or "expected", given the individual's context. This factor describes individuals constraining and minimizing the energy that they consume on a daily basis - counting calories, weighing food, and avoiding meals. Their goal is to decrease their energy consumption to as little as possible.

Next, the diagnostic criteria includes an "intense fear of gaining weight or of becoming fat, or persistent behavior that interferes with weight gain, even though at a significantly low weight" (American Psychiatric Association, 2013, p. 338). This "intense fear" goes beyond normative weight concerns, or appropriate concerns if the individual is at such a weight that gaining more would be a health issue. The intensity associated with any weight gain or "becoming fat" is clinically significant, influencing the individual's daily life, activities, relationships, and emotional wellbeing. The final criterion for a diagnosis of anorexia nervosa is a "disturbance in the way in which one's body weight or shape is experienced, undue influence of body weight or shape on self-evaluation, or persistent lack of recognition of the seriousness of the current low body weight" (American Psychiatric Association, 2013, p. 339).

Clinical Features & Presentation

Diagnostic criteria aside, certain aspects of anorexia nervosa are very apparent in clinical cases of the disorder. Specifically, patients' presentations often include overarching themes related to control and body image.

Control

One of the most challenging issues related to facing, struggling with, and overcoming anorexia nervosa is that of control. Anorexia nervosa is often described as a disorder centered around control (American Psychiatric Association, 2013), and several aspects of the disorder support this connection. For instance, the age of onset of anorexia nervosa is typically late adolescence or early adulthood, a phase of life that involves a shifting of control over one's own life. Adolescence is a period of life associated with independence, rebellion, and ambivalence toward adult authority figures (Holmbeck, 1994). Young adults or adolescents suffering from anorexia nervosa are at a place in their lives of vulnerability toward issues of control, furthering the impact of a shift in control on their identity development. As a disorder, anorexia nervosa can often be understood as being about reclaiming control of one particular aspect of one's life, controlling the physical body.

Further, around this age, individuals are often experiencing major life changes, such as moving away from their family of origin, starting college, and living independently for the first time (Holmbeck, 1994). These life transitions represent significant issues of control in one's life, as responsibilities are shifting from resting with parent or guardian figures, to the adolescent him or herself. These life transitions can elicit a variety of reactions from individuals, including regression among some people, leading them to search for ways to either give up or regain control over aspects of their lives.

Another relevant component to anorexia nervosa having to do with control includes its comorbidity with obsessive compulsive disorders. Anorexia nervosa is often diagnosed in individuals who suffer from obsessive compulsive disorders, and vice versa (American Psychiatric Association, 2013), which creates an interesting dynamic between the two. Disorders of obsessive compulsive nature include powerful, persistent, intrusive thoughts and specific, prescribed behaviors that individuals feel they must do (American Psychiatric Association, 2013). The issue of control is clearly related to this group of disorders as well, as individuals suffering from obsessive compulsive disorders simultaneously feel a need to control all aspects of their lives in order to satisfy their obsessions and compulsions, yet they also feel a loss of control to the disorder itself.

Finally, as outlined by the diagnostic criteria, individuals experiencing anorexia nervosa demonstrate an obsession with food (American Psychiatric Association, 2013). They often feel the need to control all aspects of their relationship with food, controlling intake and output of energy and meticulously counting calories. They feel the need to control every aspect of what goes into their bodies, and losing some of their ability to do so can be debilitating for someone suffering from this disorder, which can be seen when individuals with anorexia nervosa avoid eating out at restaurants, where they lose some of that control.

Body Image

Body image issues represent another overarching theme seen in the clinical presentation of anorexia nervosa. Included in the diagnostic criteria are an intense fear of becoming fat and a disturbance in the perception of one's own weight or shape (American Psychiatric Association, 2013). These aspects are necessary for a diagnosis of anorexia nervosa, and they represent significant issues with body image. An intense fear of becoming fat implies emotional connection to the way one looks, or at least the way one feels that she looks to the rest of the world, and the term fat, although undefined, suggests a worry or concern about the potential that lies there, as well as the inherent value of a person being related to their status as fat or not fat.

Another diagnostic criterion, a distortion in the individual's self-perceived weight or shape (American Psychiatric Association,

2013), also demonstrates body image issues. This factor really encompasses what body image issues are by definition, and the fact that it is included as a requisite for a diagnosis of anorexia nervosa demonstrates how deeply embedded these issues are in the disorder. A distortion in one's self-perceived body image means that the individual may see a version of him or herself in the mirror that is different from what others see. The implications of such significant body image issues run deep in individuals suffering from anorexia nervosa.

Finally, although not specifically included in the diagnostic criteria, another common clinical feature of anorexia nervosa is an obsession with the body and how others perceive one's body. This factor is closely tied with those mentioned above, but it carries enough weight to stand alone as significant. As mentioned above, this disorder often carries with it obsessive tendencies, and not only are sufferers often obsessed with the notion and experience of food and eating, but they are often also obsessed with the way they are perceived by others, as evidenced by their fear of becoming fat. This aspect of the disorder further contributes to the themes of body image issues that run throughout anorexia nervosa as a disorder.

I recognize that anorexia nervosa is a very complicated disorder and presents itself differently in different individuals, just as any psychological disorder does. The above description of common features seen in the disorder's clinical presentation is not meant to be exhaustive or applicable to every case. However, it serves the purpose of highlighting some of the major themes seen in the most typical presentations of anorexia nervosa.

Sociocultural Perspective

With its unique status as a disorder that disproportionately affects a specific demographic of individuals, as well as one that has grown substantially in prevalence over the past several decades, it seems appropriate to view the disorder through a sociocultural lens. This perspective recognizes societal and cultural factors that influence everything from individual learning and interactions, to larger phenomena and historical trends (Vygotsky, 1978). Societal factors such as wealth distribution, type of government, demographic statistics, and family structures are all incorporated in this perspective, and they combine with cultural factors, such as the role of the media, gender roles, language, and race and ethnicity, to create complex dynamics among groups of people. In this case, the sociocultural perspective is being applied to anorexia nervosa.

Cultural Obsession with the Female Body

One of the specific sociocultural factors that seems relevant to the development of anorexia nervosa is the cultural obsession with body image that exists within the United States and other Western cultures. Specifically, our culture has become increasingly obsessed with the female body (Cheney, 2011). This obsession is evidenced by advertising, other forms of media, the diet and cosmetic industries, and the language with which we talk about women, to name a few. Women in our society feel a pressure to meet the ideal that has been set for them. Female thinness has been idealized, with the message of "fat is ugly, thin is beautiful" being what most people are receiving from our society, particularly when it comes to the female body (Malson, 1998).

Looking at the fashion industry alone, this message rings loud and clear. The average model size has decreased dramatically over the past century (Sypeck, Gray, & Ahrens, 2004), setting an increasingly unrealistic standard for young girls and women to strive toward. Historically, those fashion models and celebrities who were seen as beauty ideals tended to be larger than those we see today, and many of these models would be considered "plus-sized" by today's standards. What developing girls are left with is the message that thinner is better.

Furthering this issue, our technological abilities have also increased dramatically over the years. Now photo editors have the ability to decrease a model's waist measurements, erase curves, and significantly alter the image that is presented to consumers of what a model actually looks like. With software programs like Photoshop, there can exist a huge discrepancy between how the model looked at the time of the photo shoot and the image that is released to the public, even changing factors such as eye or hair color, lighting, and facial expression (Bianco, 2014). Although a great deal of attention has been called upon the use of such techniques in media and advertising in recent years, the images we see in magazines and other sources are still drastic modifications of the original image, and it can be easy to forget this when we see the images in our everyday lives. When an adolescent girl sees a model in an advertisement, she is probably not thinking about the ways in which the image was manipulated, but rather, the ways in which the image differs from how she sees herself.

The cosmetics industry also contributes to this phenomenon. From makeup to surgery, we are told by today's society that if we are unhappy with the way we look, there is a remedy for it (Calogero, Pina, Park, & Rahemtulla, 2010). As girls and women are striving to meet the unrealistic standards of beauty set forth by the fashion and advertising industries, this sends the message that not only are there things you should dislike about your body, but also that you should want to change them and that you are able to do so with the right tools. Furthering this disconnect from reality, this puts more pressure on women to control what they look like.

Finally, the objectification of women's bodies is another massive contribution to the way in which our culture addresses women's body image (Calogero et al., 2010). Media and advertising again are a major source of this factor. Advertisements use female models to not only advertise clothing or cosmetic products intended for women's use, but for everything from cigarettes to phone plans to men's cologne, etc. Further, when female bodies are used in advertising, they are often depicted as commodities that men desire or are owed, removing women's sense of agency, which relates to a second sociocultural influence on the development of anorexia nervosa.

Women's Historical Struggle with Control

Another sociocultural factor that seems particularly relevant to anorexia nervosa as a disorder is women's historical struggle for control over their own lives and bodies. From the right to vote to the right to make decisions about one's own reproductive health, women have been fighting to overcome their "social powerlessness" and gain control over their own lives for centuries (MacSween, 1991, p. 64).

Women have had to fight for the right to make basic, foundational decisions about their own lives. Gaining the right to vote, the right to receive an education, and the right to work are just a few of the battles women have faced, and even more battles continue today, such as the right to receive equal pay or benefits (Orloff, 1993).

Particularly crucial to this conversation, women have also had to work to gain control over their own bodies. From health and reproductive rights, to issues of sexual assault and violence, women are still fighting this battle today. Women are not automatically granted the right to control decisions that are made about their own bodies, such as for issues like abortion and birth control. Further, related to the discussion above that women's bodies have been objectified by our society, the female body is often seen as an object that others can take as their own property. This issue becomes increasingly dangerous as instances of sexual assault and violence against women feel as though they are becoming more commonplace and the norm.

As gender roles in our society continue to shift, women are also experiencing issues of control related to decisions about working, raising families, and living independently. There still exists a stigma within our society attached to the idea of women working while also having a family, or in place of having a family, due to the generalizations about women's appropriate place in society.

Conclusions & Implications

I argue here that the above sociocultural factors, namely our cultural obsession with the female body and women's historical struggle to gain control over their own lives and bodies, combine in a unique way that sets the framework and contributes a significant risk factor for the development of anorexia nervosa. This unique combination sets the stage for adolescent and young women to become susceptible to developing anorexia nervosa, as they try to navigate the issues of control and body image. This is reflected in the fluctuating prevalence rates of the disorder, corresponding to shifting gender roles and sociocultural dynamics (Perlick & Silverstein, 1994). Of course, there are other factors at play in the development of anorexia nervosa, just as there are with any psychological disorder. Risk factors such as genetic predisposition, family history and family dynamics, socioeconomic status, a history of trauma, etc., all interact to contribute to whether or not an individual develops anorexia nervosa. However, as the relationships presented here suggest, as a society and culture, we need to reconsider the ways in which we address mental health issues, putting more weight on sociocultural influences where it is necessary.

Caution should be taken, however, as such an approach could lead to blaming all mental health disorders on societal and/or cultural factors, which does nothing for the individual suffering. Rather, as clinicians, researchers, and individuals, we need to address psychological disorders on an individual basis, while also striving toward larger, structural, meaningful changes as a society and a culture.

References

American Psychiatric Association. (2013). Diagnostic statistical manual of mental disorders. Arlington, VA: American Psychiatric Publishing.

Bemporad, J. R. (1996). Self-starvation through the ages: Reflections on the pre-history of anorexia nervosa. *International Journal of Eating Disorders, 19*(3), 217-237.

Bianco, A. (2014). Pervasive unreality: Reining in Photoshop. *Sprinkle: An Undergraduate Journal of Feminist and Queer Studies, 7*, 75-81.

Calogero, R. M., Pina, A. Park, L. E., & Rahemtulla, Z. (2010). Objectification theory predicts college women's attitudes toward cosmetic surgery. *Sex Roles,* 63(1-2), 32-41.

Cheney, A. M. (2011). "Most girls want to be skinny": Body (dis)satisfaction among ethnically diverse women. *Qualitative Health Research, 21*(10), 1347-1359.

Fallon, P., Katzman, M.A., & Wooley, S.C. (1996). Feminist perspectives on eating disorders. New York, NY: The Guilford Press.

Hepworth, J. (1999). The social construction of anorexia nervosa. London: Sage Publications.

Hesse-Biber, S., Leavy, P., Quinn, C.E., & Zoino, J. (2006). The mass marketing of disordered eating and eating disorders: The social psychology of women, thinness and culture. Women's Studies International Forum, 29, 208-224.

Holmbeck, G. N. (1994). Adolescence. In *Encyclopedia of Mental Health* (Vol. 1, p 1-12). San Diego, CA: Academic Press.

Keel, P. (2010). Epidemiology and course of eating disorders (p. 25-32). In W.S. Agras (Ed.) *The Oxford handbook of eating disorders*. New York, NY: Oxford University Press.

Kilbourne, J. (1994). Still killing us softly: Advertising and the obsesssion with thinness (p. 395-418). In P. Fallon, M. A. Katzman, & S. C. Wooley (Eds.) *Feminist perspectives on eating disorders*. New, York, NY: The Guilford Press.

Levine, M., & Smolak, L. (2010). Cultural influences on body image and the eating disorders. In W.S. Agras (Ed.) The Oxford handbook of eating disorders (pp. 223-246). New York, NY: Oxford University Press.

MacSween, M. (1993). *Anorexic bodies: A feminist and sociological perspective on anorexia nervosa*. New York, NY: Routledge.

Malson, H. (1998). *The thin woman: Feminism, post-structuralism, and the social psychology of anorexia nervosa*. London: Routledge.

Orloff, A. S. (1993). Gender and the social rights of citizenship: The comparative analysis of gender relations and welfare states. *American Sociological Review, 58*(3), p. 303-328.

Perlick, D. & Silverstein, B. (1994). Faces of female discontent: Depression, disordered eating, and changing gender roles (p. 77-93). In P. Fallon, M. A. Katzman, & S. C. Wooley (Eds.) *Feminist perspectives on eating disorders*. New, York, NY: The Guilford Press.

Sypeck, M. F., Gray, J. J., & Ahrens, A. H. (2004). No longer just a pretty face: Fashion magazines' depictions of ideal female beauty from 1959 to 1999. *International Journal of Eating Disorders, 36*(3), 342-347.

Vygotsky, L. S. (1978). *Mind and society: The development of higher psychological processes*. Cambridge, MA: Harvard University Press.

Anorexia Nervosa in Young Women and DSMV

Nosheen Siddiqui

When Anorexia Nervosa is mentioned stereotypical images of frail, boney, stick figure, unhealthy, self-obsessed woman come to mind. The Diagnostic and Statistical Manual of Mental Disorders V says: "Feeding & eating disorders are characterized by a persistent disturbance of eating or eating related behavior that results in the altered consumption or absorption of food and that significantly impairs physical health or psychosocial functioning" (DSM V, 2013, p. 329).

Anorexia Nervosa diagnostic criteria are as follows in the DSM V: Restriction of energy intake relative to requirements, leading to significantly low body weight in the context of age, sex, developmental trajectory, and physical health. Significantly low weight is defined as a weight that is less than minimally normal or, for children and adolescents, less than that minimally expected. B. Intense fear of gaining weight or becoming fat, or persistent behavior that interferes with weight gain, even though at a significantly low weight. C. Disturbance in the way in which one's body weight or shape is experienced, undue influence of body weight or shape on self-evaluation, or persistent lack of recognition of the seriousness of the current low body weight (DSM V, 2013, p.338-339).

To summarize the diagnostic criteria, it is clear that Anorexia Nervosa is characterized by self-starvation and weight loss. The individual starves herself because she has a fear of gaining weight. The person correlates food with fat and doesn't realize the seriousness of the disorder. Health risks vary and can include osteoporosis, amenorrhea, dehydration, renal failure, dental decay, slow heart rate and low blood pressure can cause heart failure (National Eating

Disorders, 2014). There are two subtypes of Anorexia Nervosa. The first is the restricting type in which the person controls and counts energy intake, exercises obsessively, dieting or fasting to lose weight in three months' time. The second is the binge-eating purging type of Anorexia in which a person intakes energy but purges the food to lose weight in three months' time. Often times self-induced vomiting, diuretics, enemas and laxatives are used to purge the food and prevent weight gain. According to the DSM V, the onset of the disorder typically begins in adolescence, and it is clinically reported more commonly amongst young women than men. The ratio for females to males is said to be 10:1. It is known to develop after what may be considered a stressful time in an individual's life (DSM V, 2013, pg. 339, 341).

Anorexia Nervosa at the superficial level may seem to be only about preoccupation with food and weight loss but that is often times not the case. There are usually deeper issues that may be the reason for the disorder. Anorexia is used as a coping mechanism. Those with Anorexia Nervosa use food and the control of it to cope with feelings and problems in life which seem out of their control. Controlling the intake of food and losing weight, they seem to gain a sense of control over their emotions and most importantly their lives. The stressors causing the disorder may be a disturbance in the individuals' psychological, interpersonal, social or biological factors (NEDA, 2014).

Feminist View

Feminist viewpoints focus on gender equality politically, economically and socially for women. During what is called the second wave of feminism in 1968 a protest made headlines. The Miss America Pageant protest was led by a feminist activist by the name of Robin Morgan and organized by a group called New York Radical Women. In addition to being an activist Robin Morgan was also a writer and editor and on behalf of NYRW; she wrote the agenda for the protest. She made clear that the beauty standards were ridiculous, unattainable and also racist since there were no African American women as finalist in the Miss America Pageant. On the day of the Miss America Pageant, September 7, 1968, nearly four hundred people protested against it. At the protest, protesters made their point by throwing girdles, heels, bras, and magazines like *Playboy* and *Ladies Home Journal* in a trash can they called the freedom trash can. There was a broader media coverage along with television viewers that year than the previous years of Miss America Pageant. Coverage for the protest got a lot of attention which proved to be beneficial to the Women's Rights Movement. ("Miss America", 1999-2001).

Objectification theory by Fredrickson & Roberts supports that women are treated and seen as objects given recognition as a physical body rather than as a person. The theory suggests that sexual objectification of women is linked to mental health issues in women by what they refer to as the direct and indirect manner causing eating disorders, sexual dysfunction and depression. The direct way is referred to as the external in which women experience the sexual objectification. The second manner is the indirect when women internalize the sexual objectification. The internalization alters the way they perceive their own identity by viewing themselves as an object valued by appearance rather than competence. This causes anxiety and depression. A review of studies conducted by the American Psychological Association in 2007 reflected that the media exposure of women in commercials, music lyrics, music videos, movies, and magazine advertisements projected them in a more sexual and objectifying way than men. Women were presented in a manner which emphasized their bodies in revealing outfits, and they were treated as objects. (Szymanski, Moffitt, & Carr, 2011).

Feminist view & DSM V

If we apply the feminist perspective to the DSM V we can see that the DSM fails to emphasize Anorexia Nervosa as socially caused. Rather the DSM V's criteria focus solely on the individual and the obsession to control food intake to prevent weight gain. For example DSM V criteria states B. Intense fear of gaining weight or becoming fat, or persistent behavior that interferes with weight gain, even though at a significantly low weight. Also, criteria C. Disturbance in the way in which one's body weight or shape is experienced, undue influence of body weight or shape on self-evaluation, or persistent lack of recognition of the seriousness of the current low body weight (DSM V, 2013).

My questions is, where is this intense fear of gaining weight or becoming "fat" coming from? In my opinion the answer could be low self-esteem which typically occurs when one compares themselves against someone they might view as superior. I know the cause of any eating disorder is far more complex, and each case is as unique as the individual but I think DSM should shed more light on factors pertaining to social media, magazines, celebrities, television, and troubled interpersonal relationships instead of as the problem within the person. The diagnosis of anorexia in the DSM V criteria puts the burden of shame and embarrassment on individuals diagnosed by focusing on individual pathology.

A study done by Anne Becker in 2002 called *Fijian Islands: How Television Changed the Cultural Landscape* is a prime example of the influence television has on young women and body image. Before television was introduced to Fijian culture in 1995 body image and dieting were nonexistent. Instead a good appetite and weight size were accepted and appreciated without criticism. Three years later in 1998 when data were gathered, there was a shift in the findings of young Fijian women in relation to body image. Data revealed that about 80% percent of Fijian girls were influenced by Western television and weight loss. The young women associated weight loss and body image with better career opportunities. Also, self-induced vomiting was at zero percent in 1995 and after three years of television influence in 1998 it was at 11.3% (Costin, 2007).

Findings from a study called "Shaping the Effects of Television on Adolescents' Body Image Disturbance" support a similar idea to the study done by Anne Becker. The data collected in this research indicated that young women who view television happen to be dissatisfied with their weight, body size and exhibited eating disorder symptoms more often than those who viewed less television. The study further examined parent involvement and communication styles. It was found that parents played an important role in how young women processed the images viewed on television. Positive open communication played an integral role by parental discouragement in behavior and thoughts related to eating disorders (Botta & Nathanson, 2003).

Both studies support that television has affected young women negatively. It is obvious by the studies that television upholds a negative image of what an average woman should look and feel like. In addition, if more women are influenced by television that could mean one thing: that the average majority are not the size of women who appear on television. Not all young women are aware about behind the scenes of television preparations such as the hours of heavy makeup, hair styling, perfect lighting, and girdles used to make actors look a certain way. I remember as a kid I once heard Oprah Winfrey from the talk show "Oprah" mention that it took a village for her to get ready for the show. She was obviously referring to her wardrobe, hair, and makeup stylist efforts in helping her maintain a certain image.

Celebrity singer Demi Lovato has suffered from Anorexia Nervosa. Demi has come forth publicly about her suffering from the disorder. She has spoken about anorexia nervosa and her struggle to be a certain size when she started her career as a singer. She mentioned in her video interview that she had compared her body size to others and in pursuit of being thin she developed the disorder. After treatment and overcoming anorexia, bulimia, and self-mutilation she has now embraced her body size. She stated in the video interview, "I'm heavier than I use to be but I glow now and take care of myself" (Nightline, 2013).

I think it's remarkable that some celebrities like Demi Lovato are coming forth with their struggles with weight gain and growing up with the pressures to be thin. By disclosing her illness she has broken down the barrier between herself and the millions of girls idolizing her. She has normalized her image as that of an average young women in America with struggles. Since celebrities' voices are heard worldwide through media, it is better they are followed for positive reinforcements rather than just the false façade of being what is considered "perfect" or in other words, thin. Her story can encourage many more young women around the world to embrace themselves for who they are and not who they should be. This might prevent eating disorders or help those suffering to overcome them. I think Demi Lovato has set a good example for young women by coming forth with her struggles with eating disorders.

I came across this poem by Maya Angelou (1978) that I found to be relevant to body image and self-worth based on both the inner and external qualities:

Phenomenal Woman

Pretty women wonder where my secret lies.

I'm not cute or built to suit a fashion model's size

But when I start to tell them,

They think I'm telling lies.
I say,
It's in the reach of my arms,
The span of my hips,
The stride of my step,
The curl of my lips.
I'm a woman
Phenomenally.
Phenomenal woman,
That's me.

I walk into a room
Just as cool as you please,
And to a man,
The fellows stand or
Fall down on their knees.
Then they swarm around me,
A hive of honey bees.
I say,
It's the fire in my eyes,
And the flash of my teeth,
The swing in my waist,
And the joy in my feet.
I'm a woman
Phenomenally.

Phenomenal woman,

That's me.

Men themselves have wondered

What they see in me.

They try so much

But they can't touch

My inner mystery.

When I try to show them,

They say they still can't see.

I say,

It's in the arch of my back,

The sun of my smile,

The ride of my breasts,

The grace of my style.

I'm a woman

Phenomenally.

Phenomenal woman,

That's me.

Now you understand

Just why my head's not bowed.

I don't shout or jump about

Or have to talk real loud.

When you see me passing,

It ought to make you proud.

I say,

It's in the click of my heels,

The bend of my hair,

The palm of my hand,

The need for my care.

Cause I'm a woman

Phenomenally.

I think this poem by Maya Angelou is a depiction of a strong woman who exudes confidence and acceptance of herself with great conviction. The poem is a great example of how young women should view and value themselves. She begins the poem by addressing the "pretty women" of her secrets of being the "phenomenal woman". She sets herself apart from the cute model size and goes on to describe the span of her hips and the stride of her step. She takes pride in her body size but goes on to address men are intrigued by her inner mystery making clear that her self-worth isn't dependent on just her physical being but also her inner strengths and qualities.

Social Media such as Facebook has a huge impact on body image. Kelsey Hibberd now a twenty year old young woman from Southend recalls her teenage years. She mentioned that she had been teased for being tall and pudgy. In return she had become self-conscious about her body and appearance. Her Facebook friends list was comprised of those she knew who would not pick on her. She mentioned if she didn't keep her friends list minimal then she would definitely experience far more teasing regarding her looks (Roxby, 2014).

Similarly in other health news, a reporter covered 'Selfie' to be one reason for an increase in dissatisfaction with body image in women. 'Selfie' defined in Merrium-Webster dictionary is an image of oneself taken by oneself using a digital camera especially for posting on social networks. A research study surveyed 861 female college students conducted by Ohio University, Iowa University, and University of Strathclyde found a connection between the amount of time spent on social media networks and body image. Women who spent more time on Facebook were found to compare themselves to their friends and developed a negative body image. A particular distinction was made by Petya Eckler from the University of Strathclyde in Glasgow, reporting that " Spending more time on Facebook is not connected to developing a bad relationship with food but there is a connection to poor body image" (Briggs, 2014).

Growing up in the era of social networks like Facebook, Myspace, Instagram and Twitter, I can speak from personal experience regarding the usage and observations I have made about social networking sites. I'm not going to deny the fact that these social networks do not serve a positive purpose because like any other thing they too have both pros and cons. Yes, I personally think these social networking websites make it easier to stay in touch with old friends and family and that is a good thing.

At the same time a question comes to my mind, how much of being connected to people virtually is healthy? I observed friends, family and acquaintances on these social networking websites to be in a pursuit to display the best image of themselves socially, intellectually and more so physically. There is a great emphasis on the physical body by being preoccupied with appearance because now we are connected to friends, family and acquaintances who we otherwise may not meet in person, so it's an additional pressure to put the best image forth by attempting to appear attractive. I found it to be a constant cycle of people in need of self-approval by hunting for compliments by their followers and friends on social networking sites. In my observation, attention-seeking people would change profile pictures often to what they consider to be more attractive than their last photo.

Pictures on Instagram for example are filtered and made to seem flawless since the feature of the application offers a variety of filters and cropping options to choose from. What message is this sending to people and especially young women? To me the message is to cover up reality and hide behind that which isn't real. It's not surprising to me that women develop a negative body image and eating disorders because they are constantly comparing themselves to best edited versions of friends, family and celebrities on social media.

I have seen many people feed their ego and self-worth from the number of likes and comments they receive. I remember seeing a few comments on Facebook photos as such: "you look so skinny", "you look pretty and slim", or "have you been working out?" All three comments have two things in common 1. They're referring to the superficial aspect of an individual disregarding their personal characteristics and 2. The comments insinuate a weight loss thus indicating now the person in question looks better than they once did. Thus, I agree with the health report discussed earlier regarding the link between time spent on social networking sites like Facebook and internalizing a negative body image in women.

Conclusion

In my clinical judgment, I have come to the conclusion that our society plays a huge role in the development of eating disorders in young girls and women. I acknowledge the fact that Anorexia Nervosa and other eating disorders are far more complex than to simply focus on the individual as the sole cause of the problem. As a future social worker, I can't accept that this disorder simply lies within the person. A deeper introspection needs to be made in order to come close to understanding the pathology of the disorder. As a social worker I'm naturally inclined to not only examine the individual but also the environment the person belongs to.

In the time we live in television, social media, magazines, advertisements and celebrities project a false standard of what is deemed physically attractive or acceptable, in this case it is being "thin". In my opinion, the youth today need more mentors, better role models, actively involved family members who can help them develop strong personalities by having self-worth based on inner qualities rather than just physical. They also need to develop resiliency skills to become better at overcoming hardships in order to live healthier and more fulfilled lives. Therefore, DSM V should emphasize its focus on environmental factors by broadening the diagnostic criteria rather than only focusing on individual pathology. Focus on the individual can lead to shame in an already stigmatized disorder. By shifting focus in DSM V criteria for the diagnosis of anorexia nervosa from the individual to the environment can perhaps help reduce the stigma and promote a healthy body image that can possibly lead to prevention of anorexia nervosa and other eating disorders.

References

American Psychiatric Association. (2013). *Diagnostic and statistical manual of mental disorders*: DSM-5. Washington, D.C: American Psychiatric Association.

Angelou, Maya (1978). Phenomenal Woman. *Poetry Foundation*. Retrieved from https://www.poetryfoundation.org/poem/178942.

Briggs, Helen (2014, April). 'Selfie' body-image warning issued. BBC News Health. Retrieved from https://www.bbc.com/news/health-26952394.

Botta, Renee A. & Nathanson, Amy I. (2003). Shaping the effects of television on adolescents' body image disturbance. *Communication Research, 30 (3)*, 304-331

Costin, Carolyn (2007). The eating disorder sourcebook, 3rd edition. New York, NY: Mc Graw Hill.

Miss America (1999-2001). People & events: The 1968 protest. Retrieved from PBS Online
http://www.pbs.org/wgbh/amex/missamerica/peopleevents/e_feminist.html

National Eating Disorder Association (2014). Anorexia Nervosa. Retrieved from https://www.nationaleatingdisorders.org/anorexia-nervosahttps://www.nationaleatingdisorders.org/anorexia-nervosa

National Eating Disorder Association (2014). Anorexia Nervosa. Retrieved from https://www.nationaleatingdisorders.org/factors-maycontribute-eating-disorders.

Nightline (2013, May, 30). Demi Lovato talks about eating disorder. [YouTube] Retrieved from www.youtube.com/watch?v=_8030067cjU.

Roxby, Phillippa (2014, October). Does social media impact body im   age? Retrieved from http://www.bbc.com/news/health-29569473.

Selfie. 2014. In Merriam-Webster.com. Retrieved December 8, 2014, from http://www.merrium-webster.com/dictionary/selfie

Szymanski, Dawn M., Moffitt, Lauren B., & Carr, Erika R. (2011).Sexual objectification of women: Advances to theory and research. The Counseling Psychologists 39 (1) 6-38.

Conduct Disorder in Girls

Jennifer Allen

Introduction

The Diagnostic and Statistical Manual of Mental Disorders, Fifth Edition (DSM-5) is the primary tool that mental health professionals in the United States use to assist them in diagnosing various mental illnesses and disorders. It categorizes various symptoms in order to provide clinicians, psychiatrists, social workers, insurance companies and other various mental health professionals a systematic approach to diagnosing clients. Although the DSM serves as the so-called "Bible" for the profession, it still leaves much room for improvement with regards to assisting mental health professionals to provide the most comprehensive care for their clients or potential clients.

This writer will be examining the diagnostic criteria for conduct disorder, with regards to discussing how the current criteria are too narrow and stereotypical in scope to appropriately encompass the differences in pathology between men and women.

Diagnostic Criteria of Conduct Disorder

According to the DSM-5, Conduct Disorder is "a repetitive and persistent pattern of behavior in which the basic rights of others or major age-appropriate societal norms or rules are violated" (APA, 2013, p. 469). The four main maladaptive categories that are present in conduct disorder are aggression towards people and animals, destruction of property, deceitfulness or theft, and serious violations of rules. Often individuals with conduct disorder may bully and initiate physical fights with others, are physically cruel to people and animals, commit rape and robbery, start fires, con others, shoplift, break curfew and run away from home for multiple days. Although these are some common examples of symptoms for people with conduct disorder, this is not an exhaustive list.

The three subtypes of conduct disorder are childhood-onset type, adolescent-onset type, and unspecified onset type. Prior or post the age of ten is the differentiating factor, and if the age of onset is unknown then it would be classified as unspecified. Lack of remorse, lack of empathy, being unconcerned about performance and shallow or deficient affect are examples of the "with limited prosocial emotions" specifier (APA, 2013). It is important to confirm information from various reporting sources across multiple settings to confirm prior to indicating a specifier.

There are also several other associated features that help support a diagnosis of conduct disorder. For example, individuals who present with aggression often times misperceive the intentions of others as being more hostile and threatening than they are in reality, therefore these individuals feel justified by responding to these situations with aggression. The population prevalence ranges from 2%-10%, with an average of 4% (APA, 2013). Boys are diagnosed at a higher prevalence rate than girls, but prevalence rates remain the same across countries of different races and ethnicities. An individual's temperament, environment and genetic make-up are all examples of potential risk factors for conduct disorder. There are several functional consequences of conduct disorder. For example, school suspensions, expulsions, legal issues, sexually transmitted diseases, and physical injury from accidents or fights are all common problems that individuals with conduct disorder may have to deal with. In addition, individuals diagnosed with conduct disorder also commonly have other co-occurring disorders, such as oppositional defiant disorder, Attention-deficit/hyperactivity disorder, bi-polar disorder, intermittent explosive disorder and various other adjustment disorders.

Definition of Aggression

Generally speaking, the definition of aggression is a hostile or violent behavior or attitudes towards another. When referring to conduct disorder, this definition is too vague to fully understand how aggression may distinctly present itself in children of different genders. There are some key differences in how young girls exhibit aggression versus how boys exhibit aggression. According to Kann & Hanna (2000), boys display more externally directed behaviors such as stealing, lying, fighting, and destructiveness. In contrast, girls display more internally focused behaviors which include anxiety, shyness, withdrawal, hypersensitivity, and physical complaints. Another way to explain this would be to categorize behaviors that boys display as "overt aggression" and girls' behaviors would be described as "relational aggression." The DSM-5 only briefly touches on this topic under the heading of gender-related diagnostic issues. However, this writer feels that more attention needs to be given to this distinct difference in diagnosing criteria. Both forms of aggression, whether overt/physical or covert/relational, are equally important when diagnosing conduct disorder in children and adolescents.

An additional barrier when recognizing the various forms of aggression in females is the concept that girls have more sophisticated social skills compared to their male classmates. This social awareness makes it easier for girls to conceal their deceitfulness and escape detection longer (Atkin-Little, Delligatti & Little, 2003). With regards to school settings, many school personnel are more focused on the physical and overt aspects of aggression that boys are more likely to display. For example, fighting in the hallways, damaging school property, and overtly breaking school rules is what tends to raise red flags for adults in recognizing that there's a problem with a particular student.

"Girls with conduct disorder, however, may be more likely to practice covert relational aggression, through behaviors intended to significantly damage another child's friendships or feelings of inclusion by the peer group, and other covert forms of deceitfulness and theft that may not be as apparent to school staff" (Atkin-Little, Delligatti & Little, 2003, p. 185). Basically, girls display relational aggression that focuses on spreading gossip to get other peers to reject a child, withdrawing friendship, or excluding them from the group in order to intentionally hurt or control another peer. Since these behaviors are not as easily detected by adults and school personnel, it is essential for teachers, school social workers and parents to talk with other peers in order to fully assess the extent of the aggression or problem. What this writer found interesting related to aggression, is that girls and boys both choose to use the form of aggression that is intended to damage their peers' most valued goals, physical dominance in boys and social relationships in girls (Crick & Grotpeter, 1995).These valued social goals highlight the importance of social context theory and how it has affected how we have diagnosed mental disorders over time.

Limitations of Research on Conduct Disorder in Girls

There are several reasons that could be contributing to the lack of research on conduct disorder in girls. First, gender stereotypes and other various social structures in society that perpetuate those stereotypes contribute to the idea that conduct disorder among girls is very rare. Unfortunately, this is why most outcome studies only focus on boys or all children as a whole without separately examining how girls are affected by conduct disorder (Atkin-Little, Delligatti & Little, 2003).

A contributing factor to this problem is that most studies regarding conduct disorder are predominately conducted within the criminal justice system and other various in-patient settings within the psychiatric system that tend to be more male dominated. In addition, since the DSM-5 diagnostic criteria are largely derived from these types of studies, there is a significant lack of female samples in order to effectively validate these studies and how they relate to the female population.

Socialization Factors

Many factors that play a role in how our society socializes its children are apparent as early as infancy in some cases. How young girls and boys are socialized to appropriately express anger is an example of this disparity. For example, female infants are encouraged to suppress feelings of anger and probably encounter negative responses from caregivers. In contrast, male infants that express anger received a more empathic response (Atkin-Little, Delligatti & Little, 2003).

Studies have shown that caregivers are more likely to tolerate certain behaviors from children of the same gender as the caregiver. Female teachers and mothers are more likely to not see a problem with the internalizing behaviors that girls display, but are more likely to focus on boys in the classroom or in the home as being problematic for displaying more externalizing behaviors. This may be another example of why girls are being under-diagnosed. According to Kann & Hanna (2000), since girls with conduct disorder display more covert behaviors, their friends may fail to recognize a potential problem and not report it to adults. The expectations of parents, teachers, and peers all play a role in socializing what behaviors are deemed appropriate for the genders and could have a direct correlation to the disparities in how boys and girls are diagnosed with conduct disorder. In addition, many different social environments such as family structures, schools, and churches further reinforce social stereotypes related to gender.

Risk Factors

Although extensive research is limited, several risk factors have been identified to contribute to a diagnosis of conduct disorder and even some factors that are specific to females. Temperament, genetics, intellectual functioning and behavioral factors can be classified under child factors. According to Atkin-Little, Delligatti & Little (2003), Neurodevelopmental impairments such as deficits in verbal and executive functioning have been found to be indirect predictors of conduct disorder. Some examples of parent and family factors are parent psychopathology (specifically depression), parent criminal behavior, and inconsistent parenting. In addition, large family size, inconsistent supervision, and low levels of parental warmth, affection, and emotional support are other parent and family risk factors. Some school related risks factors are school environments in which there is not a strong emphasis on academic performance, school buildings that are deteriorating, and lack of teachers' ability to be handle student issues and problems.

There are certain parent and family risk factors that are specific to girls. Girls with conduct disorder had a higher percentage of mothers who were hostile in their parenting approach. A history of being a victim of sexual abuse was also a risk factor. Also, Borderline Personality disorder was more common among girls versus boys. In contrast, boys were more likely to be diagnosed with Attention Deficit/Hyperactivity Disorder and Intermittent Explosive Disorder (Atkin-Little, Delligatti & Little, 2003). Identifying these risk factors is an essential component to providing young girls the earliest intervention possible to receiving services in hopes of avoiding some of the negative consequences that can accompany individuals with conduct disorder.

Summary and Conclusions

Obviously, there is still a lot of research that needs to be conducted with regards to girls and conduct disorder. The current diagnostic criteria are based on research conducted and validated strictly on boys. More studies need to strictly focus on girls and the range of symptoms that they display. Mental Health and Education professionals need to be informed about the socialization factors that play a role in the misdiagnosing and under-diagnosing of young girls that may preventing them from early intervention and treatment that is so desperately needed. Research also needs to focus on what types of interventions will work on girls once an appropriate diagnosis is reached. One suggestion that professionals should consider is broadening the diagnostic criteria to also include more relational and covert behaviors of aggression. Hopefully, once there are appropriate diagnostic criteria to fully encompass the range of different symptoms displayed by both boys and girls more people will be able to benefit from the advantages of early intervention. Early intervention is essential to avoid negative consequences and stigmatization that can accompany conduct disorder if not appropriately addressed.

References

Akin-Little, A., N. Delligatti, & S.G., Little. (2003). Conduct disorder in girls: Diagnostic and intervention issues. Psychology in the Schools, Vol. 40(2). 183-192.

American Psychiatric Association. (2013). Diagnostic and statistical manual of mental disorders (5th ed.). Washington, DC: American Psychiatric Association

Crick, N.R., & Grotpeter, J.K. (1995). Relational aggression, gender, and social-psychological adjustment. Child Development, 66, 710-722.

Kann, T., & Hanna, F. (2000). Disruptive behavior disorder in children and adolescents: How do girls differ from boys? Journal of Counseling and Development, 78, 267-274.

Conduct Disorder: A Result of the Environment

Ashley Piotrowski

Diagnostic and Statistical Category

The Diagnostic and Statistical Manual of Mental Disorders Fifth Edition (DSM-V) is the handbook that is used by health care professionals mainly in the United States to guide the diagnosis of mental disorders (American Psychiatric Association, 2013). This handbook is meant to summarize characteristic syndromes that may point to an underlying disorder. According to the American Psychiatric Association (2013), the diagnostic criteria for conduct disorder requires that an individual exhibits

> a repetitive and persistent pattern of behavior in which the basic rights of others or major age-appropriate societal norms or rules are violated, as manifested by the presence of at least three of the following 15 criteria in the past 12 months from any of the categories… with at least one criterion present in the past 6 months. (p. 469)

The four major categories of behaviors that are manifested in various settings include: aggressive conduct that causes or threatens physical harm to other people or animals; non-aggressive conduct that causes property loss or damage; deceitfulness or theft; and serious violations of the rules (American Psychiatric Association, 2013). Usually patterns of these behaviors begin in children before the age of 13, and many youth have trouble with prosocial behaviors, such as problems with feeling and expressing empathy or remorse, and reading social cues.

Differing Theoretical Perspectives

Urie Bronfenbrenner's Ecological Systems Theory (1979) proposes that children develop within a context of their environment. There are five environmental systems in which individuals interact: Microsystem, mesosystem, exosystem, macrosystem and the chronosystem. According to Bronfenbrenner's model, individuals cannot be separated from their environmental contexts, and therefore, their behaviors can be explained by different interactions among the systems. Furthermore, systems may not directly affect the individual, but interactions among the systems and among different people within the systems can have repercussions for the individual, for example, a child's parent loses his or her job, neighborhood violence, inadequate school resources, etc.

Barbara Rogoff (2003) further explained children's development using Bronfenbrenner's model with an emphasis on cultural factors that are embedded into the different systems impacting the child's development. Human functioning occurs in a dynamic nature of culture with children growing up in a family and a community in which the child participates and learns various practices and rituals (Rogoff, 2003). Thinking and learning are both part of the cultural developmental process, including the behaviors that are culturally appropriate (Rogoff, 2003). Thus, the behaviors related to the origin and maintenance of conduct disorder entail multiple systems that are nested within one another and the culture, which all interact to influence the individual.

Paradigm Shift

Diagnosing people through the lens of the DSM-V reflects a medical model of psychopathology, which implies that there is a physical cause to mental disorders and places blame on the individual for their mental health. Arguably, conduct disorder is embedded in the environment and in the social context rather than the individual, challenging the diagnosis of the DSM-V and attributing conduct disorder behaviors to the society or environment. Humans live within an environment that supports the interconnectedness of the different ecological systems, and to understand life's transitions and different coping skills of the individual we must look at all the differing elements and understand how they influence one another.

Differing Impacts on the Lives of Youth

Community Impact

The community that an individual grows up in provides a social context to his or her development, as discussed above, and a youth's subjective environment and exposure to violence directly influences his or her mental health (Stiffman, Hadley-Ives, Elzem Johnson, & Dore, 1999). Subjective environmental experience is the youth's perspective about his or her neighborhood's violence. Youths' behaviors are therefore in relation to their perception of their community.

According to Mrug, Loosier, and Windle (2010), there are multiple ways that exposure to community violence may affect adjustment and impede normal development. First, encounters with violence are inherently stressful and may shift the development of self-regulation, including emotions, behaviors and attention. Second, witnessing or being a victim of violence undermines a child's sense of security and makes it difficult for him or her to engage in developmentally healthy activities and to establish secure relationships (Cummings & Davies, 1996). Finally, externalizing behavior problems may be due to repeated witnessing of or victimization by violence by desensitization of children to the effects of violence by modeling aggressive behavior as an acceptable and effective strategy for achieving one's goals (Mrug, Loosier, &

Windle, 2010). A review of the literature reveals that children exposed to violence experience changes in their brains, suggesting desensitization to violence (as cited in Mrug, Loosier, & Windle, 2010). Inclusively, Guerra, Huesmann, and Spindler (2003) write that "exposure to community violence has been linked to an increase in aggression, aggressive fantasies and behaviors over time" (as cited in Mrug, Loosier, & Windle, 2010, p.70).

Socioeconomic Status

Living in poverty is a stressor and risk factor for many health and mental health disorders. Poverty, in combination with other risk factors, is associated with negative outcomes for children (Yeung, Linver, & Brooks-Gunn, 2002). Forty-five percent of all children, 18 years of age and younger live in low-income families (Jiang, Ekono, & Skinner, 2014). Poverty is interrelated to many factors that influence, for example, crime, education and environmental quality. Researchers controlled for race and ethnicity and found a significant correlation between lower socioeconomic status (SES) and conduct disorder (Lahey et al., 1995).

Antisocial personality disorder that is diagnosed in adults is said to be predicted by childhood conduct disorder (American Psychiatric Association, 2013). However, researchers Lahey, Loeber, Burke & Applegate (2005), suggest that conduct disorder predicts antisocial personality disorder only in lower SES families, indicating socioeconomic status, particularly low-income, to be an influential key in predicting future antisocial behaviors.

.Access to resources, particularly school resources, is heavily influenced by geographical location of the school. On a local level taxpayers' property taxes help fund education; therefore, the wealthier the area, the greater increase in ability to collect more funds in property taxes. This is directly correlated to the schools having more resources. Poor communities lack the ability to raise funding to support local school districts, which means "children with the highest needs go to schools with the least resources, the least qualified teachers, and substandard school facilities" (New America Foundation, 2014).

Racial Differences

There are large discrepancies between individuals who identify with either White or African American races on the diagnosis of conduct disorder. African American youth experience considerably more risk factors than their White counterparts, yet they are diagnosed with conduct disorder at lower rates (VanHook, 2012). As previously discussed, lower SES is a risk factor for conduct disorder, and 62% of African American youth live in low-income households according to the National Center for Children in Poverty (Chau, Thampi, & Wight, 2010). This indicates that community and environmental factors elevate the risk of conduct disorder for African American youth. This potentially means African American youth with conduct disorder behaviors are not being referred for psychological services or interventions, but rather to the criminal justice system.

"The school-to-prison pipeline is an epidemic that plagues schools across the nation" (Amurao, 2013). Particularly African American males are treated in school systems as they are in the larger society. There are large inequalities and discrepancies of discipline policies against African American males, which make them a target to become forced out of school and become stigmatized. This tends to begin a cycle of truancy for student's whose behavior, which is surface level, is actually due to underlying negative home or neighborhood environments.

Gender Differences

The ways in which individuals define and experience their worlds are influenced by gender (DeHart, Sroufe, & Cooper, 2004). Gender is socially constructed and tends to be consistent with the adult roles of the women and men in many communities around the world (Rogoff, 2003). Because of this children are learning the stereotypical behavior and characteristics for what it means for them to be male or female (DeHart, Sroufe, & Cooper, 2004). Children are socialized differently depending on their sex, and in many cultures aggression is attributed to males and nurturance to females (Rogoff, 2003). Additionally, when parents converse with their children, they tend to talk more about social events and emotions with girls than with boys, which influences girls to talk more about their emotions and recall more emotional aspects of events compared to boys (DeHart, Sroufe, & Cooper, 2004).

In regards to gender and conduct disorder, according to the DSM-V, there is a difference in the prevalence of conduct disorder in males and females. Males tend to have a slightly higher prevalence rate than females due to boys' tendency to act out violently (externalizing factors), while girls have a tendency to act out in their interpersonal relationships (American Psychiatric Association, 2013), but since gender roles are culturally socialized, males might be acting according to society's roles for male behavior.

Parental Influence

Parents have a large influence on their children. Children grow up modeling behaviors that they are exposed to within their home and community environment. Bandura's Social Learning Theory (1977) suggests that people learn from one another through modeling, imitation and observation, which would support the idea that human behavior is reciprocal between children and their parents. If children are exposed to aggression, destruction, deceitfulness, and serious violations of rules then those are the behaviors they learn to be socially acceptable. This is not to solely blame parents for children's misbehaviors, but to simply point out that for behaviors to be existent they must be learned from somewhere or someone.

Stressors in parents' lives also have effects on their children. Biological parents, in a longitudinal study of children with conduct disorder, exhibited higher rates of psychopathology (Lahey et al., 1988). Antisocial personality disorder was more prevalent in both mothers and fathers of children with conduct disorder. Substance abuse was also prevalent among fathers of children with conduct disorder. This potentially raises the question of what comes first, "the chicken or the egg". Maybe the child with conduct disorder influenced the parent's pathology; however, other studies suggest that there is an increase in children's mental disorders, including conduct disorder, particularly related to parent's substance use disorders, implying parents affect their children (Clark, Cornelius, Wood, & Vanyukov, 2004). Additionally, "analogous to their children, fathers with substance use disorders often have childhood histories of conduct disorder..." (Clark, Cornelius, Wood, & Vanyukov, 2004, p. 685). This implies a cyclical relation between parents' behaviors and mental disorders and their child's behaviors and mental disorders with potential intergenerational transmission.

Peer Influence

During adolescence, youth become highly involved with their peer relationships. During this period of development individuals may succumb to either positive or negative influences from their peers, known as peer pressure. Sometimes friends are protective factors against risky behaviors, and other times they are negatively influencing behaviors.

If youth perceive themselves as being "bad" they are influenced to act out and fit within that stereotype, living up to the stigma that has already been labeled upon them. This becomes a snowball effect for individuals diagnosed with conduct disorder building and continuing negative interaction patterns with peers, teachers, and family members (Coie & Jacobs, 1993). Peers also reinforce aggressive behavior when they succumb to the child making threats and physical force to achieve personal goals and allowing them to succeed (Coie, Dodge, Terry & Wright, 1991). Children essentially become trapped in a cycle of aggressive behavior because their actions have immediate rewards. Therefore, the school and peer environment plays an important role in promoting and maintaining maladaptive behaviors (Coie & Jacobs, 1993).

Cultural Factors in the Expression of Emotions and Behaviors

In the discussion of youths' behaviors there is an important element of culturally appropriate expression of emotions in relation to behaviors that needs to be taken into consideration. Previously there has been research in regards to cultural variations in psychopathology, but there is a lack of research in investigating the cultural symptomology of disruptive behaviors. Humans are culturally bound and we express ourselves in accordance with the cultural norms and regularities in the communities where we were raised (Rogoff, 2003). Ways of managing expressions of anger are taught to young children through cultural norms. Cultures have different ways of handling anger in regards to child-rearing patterns and children's expression of anger (Stearns, 2008). Moreover, cultures vary in their attitudes towards anger and their beliefs about the normalcy and extent to which anger can be displayed (Stearns, 2008), which directly correlates to culturally acceptable behaviors.

Although emotions are biologically programmed, the way they are displayed and perceived depends on cultural factors (Matsumoto, 1989). Additionally, the language used to describe the behaviors and emotions may translate differently from the origin of the language used for diagnosing conduct disorder. As cited in Elfenbein and Ambday (2002), Matsumoto and Assar in 1992 "proposed that the vocabulary of some languages might be better equipped to express emotional concepts than those of other languages" (p. 204). Emotion is expressed not only through the vocabulary but also through the rhythm and intonations of voices, which allows cultural group members an advantage of distinguishing between vocal expressions (Scherer, Banse, & Wallbott, 2001). This would also indicate an in-group advantage to judging emotions when there is a match between the social background and culture of the target and judge (Elfenbein & Ambday, 2002). Emotions are not the same as an individual's behavior; nevertheless emotions interact with behaviors. Emotions have evolved for our benefit of judgments and decision making, which means emotions benefit, rather than impede, social decision-making (Haselton & Ketelaar, (2005). Therefore, the behaviors behind conduct disorder may be adaptive for children living in areas that are lower income and with high rates of violence for survival.

Lack of Consistency in Diagnosing Conduct Disorder

Researchers Hsieh and Kirk (2003) illustrated limitations in the diagnosis of conduct disorder by professionals in relation to the influence of social context. The researchers allowed "experienced psychiatrists to assess the presence of a mental disorder in systematically manipulated vignettes that met the diagnostic criteria for conduct disorder" (Hsieh & Kirk, 2003, p. 883). They found that when making a judgment of whether there was a mental disorder based on features of the symptoms alone psychiatrists failed to differentiate between disordered and non-disordered adolescent antisocial behavior, leading to false over-diagnoses (Hsieh & Kirk, 2003). As cited in Hsieh and Kirk (2003), context-free applications of the DSM criteria for conduct disorder may contain a bias toward false-positive diagnoses. Youth in impoverished high crime and violent communities would more than likely meet the symptom diagnostic criteria for conduct disorder; however, most psychiatrists in the U.S. would not agree that these youth have mental disorders (Hsieh & Kirk, 2003). This indicates that there is a social context factor that needs to be taken into consideration if diagnosing an individual with conduct disorder.

Conclusion

In conclusion, when assessing conduct disorder it is pivotal to examine the role of the environment including the social context and culture of the individual. This examination of conduct disorder is to shed light and question whether using the DSM diagnostic criteria can accurately capture conduct behaviors or normal adaptions of behaviors to problematic environments. Aggression may be used by youth in certain settings as a survival tool and labeling them with conduct disorder may not help the situation, specifically in regards to one's race, if a diagnosis of conduct disorder does not lead to positive interventions. Race seems to be influential in whether adolescents receive intervention or the juvenile delinquency system.

Perhaps when symptoms of conduct disorder arise the issue lies within the environment rather than the individual all together. The environment is not meeting the child's needs, to a certain extent it is not a good fit, and environmental adjustments may help the child function better in his or her environment. In particular, school settings may not be equipped to handle children with conduct disorder, or they may not have access to the resources that could benefit the environment and children. In schools, the personnel tend to focus on behaviors rather than underlying causes of the behaviors, which in my opinion does not deal with the root cause of what are deemed as truant behaviors. Perhaps children with conduct misbehaviors are manifesting the problems of the system, and looking at the diagnostic criteria of conduct disorder through a cultural environmental lens helps to see the larger issue on a systemic community level rather than issues lying within the individual.

References

American Psychiatric Association. (2013). *Diagnostic and Statistical Manual of Mental Disorders* (5th ed.). Washington, DC.

Amurao, C. (2013). *Fact sheet: How bad is the school-to-prison pipeline?* Retrieved from http://www.pbs.org/wnet/tavissmiley/tsr/education-under-arrest/school-to-prison-pipeline-fact-sheet/

Bandura, A. (1977). *Social Learning Theory*. New York: General Learning Press.

Bronfenbrenner, U. (1979). *The ecology of human development: Experiments by nature and design*. Harvard University Press: Cambridge, Massachusetts.

Chau, M., Thampi, K., & Wight, V.R. (2010). *Basic facts about low-income children, 2009: children under age 18*. Mailman School of Public Health, Columbia University, New York, New York. Retrieved from htttp://www.nccp.org/publications/pdf/text_975.pdf

Clark, D.B., Cornelius, J., Wood, D.S., & Vanyukov, M. (2004). Psychopathology risk transmission in children of parents with substance use disorders. *AM J Psychiatry*, 161 (4), 685-691.

Coie, J.D., Dodge, K.A., Terry, R. & Wright, V. (1991). The role of aggression in peer relations: An analysis of aggression episodes in boy's play groups. *Child Development*, 62, 812-826.

Coie, J.D. & Jacobs, M.R. (1993). The role of the social context in the prevention of conduct disorder. *Development and Psychopathology*, 5, 263-275.

Cummings, E.M. & Davies, P. (1996). Emotional security as a regulatory process in normal development and the development of psychopathology. *Development and Psychopathology,* 8, 123-129.

Dehart, G., Sroufe, L., & Cooper, R. (2004). *Child development: Its nature and course* (5th ed.) New York: McGraw-Hill.

Elfenbein, H.A. & Ambday, N. (2002). On the universality and cultural specificity of emotion recognition: A meta-analysis. *Psychological Bulletin*, 128 (2), 203-235.

Haselton, M.G. & Ketelaar, T. (2005). Irrational emotions or emotional wisdom? The evolutionary psychology of emotions and behaviors. In press, J. Forgas (ed.), *Hearts and minds: Affective influence on social cogntion and behavior.* (Frontiers of Social Psychology Series). New York: Psychology Press.

Hsieh, D.K & Kirk, S.A. (2003). Diagnostic consistency in assessing conduct disorder: An experiment on the effect of social context. *American Journal of Orthopsychiatry,* 74(1), 43-55.

Jiang, Y., Ekono, M., & Skinner, C. (2014). *Basic facts about low-income children, children under 18 years, 2012.* National Center for Children in Poverty. Retrieved from http://www.nccp.org/publications/pub_1089.html

Lahey, B. B., Loeber, R., Burke, J. D., & Applegate, B. (2005). Predicting future antisocial personality disorder in males from a clinical assessment in childhood. *Journal of Consulting and Clinical Psychology,* 73(3), 389-399.

Lahey, B. B., Loeber, R., Hart, E. L., Frick, P. J., Applegate, B., Zhang, Q.,... & Russo, M. F. (1995). Four-year longitudinal study of conduct disorder in boys: Patterns and predictors of persistence. *Journal of Abnormal Psychology,* 104(1), 83-93.

Lahey, B. B., Piacentini, J.C., McBurnett, K., Stone, P., Hartdaghn, S., & Hynd, G. (1988). Psychopathology in parents of children with conduct disorder and hyperactivity. *Journal of the American Academy of Child and Adolescent Psychiatry,* 27(2), 163-170.

Matsumoto, D. (1989). Cultural influences on the perception of emotion. *Journal of Cross-Cultural Psychology,* 20, 92-105.

Mrug, S. Loosier, P.S., & Windle, M. (2010). Violence exposure across multiple contexts: Individual and joint effects on adjustment. *American Journal of Orthopsychiatry,* 78 (1), 70-84.

New America Foundation (2014). *School finance federal, state and local k-12 school finance overview.* Retrieved from http://febp.newamerica.net/background-analysis/school-finance

Rogoff, B. (2003). *The cultural nature of human development* (pp. 3-36). New York: Oxford University Press.

Scherer, K.R., Banse, R. & Wallbott, H. (2001). Emotion inferences from vocal expression correlate across languages and cultures. *Journal of Cross-Cultural Psychology,* 32, 76-92.

Stearns, D. (2008). *Anger and aggression. Encyclopedia of Children and Childhood in History and Society.* Retrieved from http://www.faqs.org/childhood/A-Ar/Anger-and-Aggression.html

Stiffman, A., Hadley-Ives, E., Elze, D., Johnson, S., & Doré, P. (1999). Impact of environment on adolescent mental health and behavior: Structural equation modeling. *American Journal of Orthopsychiatry*, 69(1), 73-86.

VanHook, C.R. (2012). Racial disparity in the diagnosis of conduct disorder. *Georgia State University Undergraduate Research Awards.* Retrieved from http://scholarworks.gsu.edu/univ_lib_ura/12

Yeung, W. J., Linver, M. R., Brooks-Gunn, J. (2002). How money matters for young children's development: Parental investment and family processes. *Child Development*, 73(6), 1861-1879.

Diagnosing Alcoholism in Youth: No Going Back

Owen D. Murphy

Overview

The fifth and newest edition of the Diagnostic and Statistical Manual of Mental Disorders (DSM-5) was released in 2013. The DSM, as many are aware, is the current clinically accepted standard for diagnosing mental disorders in the United States, which drives the creation and implementation of treatment plans and, importantly, in the vast majority of cases also facilitates payment for treatment by third-party health insurance companies. For mental health disorders and treatment in America, it all starts with the DSM.

Keeping that context in mind, this article will conduct a critique of one specific diagnosis from the DSM-5, namely Alcohol Use Disorder (AUD), or as it is more commonly known, alcoholism. In particular, we will focus on the troubling ramifications of an AUD diagnosis for youths, who are by far the age group most commonly diagnosed with the disorder (APA, 2013). The DSM diagnosis for AUD lays out objective criteria including eleven different symptoms, which will be summarized in detail.

We will then examine alcoholism from the theoretical perspective of social constructionism. This will feature a basic explanation of social constructionist theory, followed by a brief history of alcoholism in American society. We will highlight the fact that the idea of alcoholism as a disease is a classic example of a social construction, with a notable lack of scientific evidence supporting any biological mechanisms behind it. Finally, we will take a look at the current treatment environment for the disorder and the overwhelming prevalence of the disease model. While treatment will not be the focus of our discussion, an understanding of the treatment environment is essential in order to grasp the harmful consequences for a person who receives the diagnosis.

Alcoholism, as it shall be illustrated, is far from the singular experience the DSM attempts to present. Behaviors and cultural norms vary widely, and it affects individuals in vastly different ways based on any number of factors including gender, ethnicity, age, sexual orientation, economic class and upbringing. In America currently, excessive alcohol consumption is especially common among youths of all ethnicities, both adolescents and young adults, particularly on college campuses. In applying the criteria in the DSM to this population, there is a very real possibility of an AUD diagnosis for someone who in fact does not experience problematic behavior, and is not likely to develop alcoholism.

On top of that, evidence shows that, contrary to public perception, most people diagnosed with AUD make a successful recovery. Many problem drinkers make a decision to reduce their drinking without a formal diagnosis, and successfully do so without formal treatment (Peele, 2004). This suggests that for many, a diagnosis is unnecessary and even counterproductive due to the intense feelings of disempowerment that are almost inescapable in the current treatment environment.

It is clear that offering specific criteria and objective symptoms in the DSM as a basis for diagnosing AUD is an oversimplification at best, and reckless at worst. This critique concludes that the DSM, by presenting a seemingly objective list of criteria for a very subjective phenomenon like alcohol addiction, facilitates the disempowerment and stigmatization of adolescents and young adults.

Diagnostic Criteria

The DSM lists the following diagnostic criteria for Alcohol Use Disorder:

A. A problematic pattern of alcohol use leading to clinically significant impairment or distress, as manifested by at least two of the following, occurring with a 12-month period:

1. Alcohol is often taken in larger amounts or over a longer period than was intended.

2. There is a persistent desire or unsuccessful efforts to cut down or control alcohol use.

3. A great deal of time is spent in activities necessary to obtain alcohol, use alcohol, or recover from its effects.

4. Craving, or a strong desire or urge to use alcohol.

5. Recurrent alcohol use resulting in a failure to fulfill major role obligations at work, school, or home.

6. Continued alcohol use despite having persistent or recurrent social or interpersonal problems caused or exacerbated by the effects of alcohol.

7. Important social, occupational, or recreational activities are given up or reduced because of alcohol use.

8. Recurrent alcohol use in situations in which it is physically hazardous.

9. Alcohol use is continued despite knowledge of having a persistent or recurrent physical or psychological problem that is likely to have been caused or exacerbated by alcohol.

10. Tolerance, as defined by either of the following:
a) A need for markedly increased amounts of alcohol to achieve intoxication or desired effect.
b) A markedly diminished effect with continued use of the same amount of alcohol.

11. Withdrawal, as manifested by either of the following:
a) The characteristic withdrawal syndrome for alcohol (also defined in DSM-5).
b) Alcohol (or a closely related substance, such as a benzodiazepine) is taken to relieve or avoid withdrawal symptoms.

Specify current severity:
  Mild: Presence of 2-3 symptoms.
  Moderate: Presence of 4-5 symptoms.
  Severe: Presence of 6 or more symptoms.

It is important to note that the DSM gives guidance on remission, with *early* remission specified if no symptoms have been met for over three months, and *sustained* remission specified if no symptoms have been met for a period of twelve months or longer. However, both provide the exception that "Criterion A4, Craving, or a strong desire or urge to use alcohol, may be met" (APA, 2013). I would point out that this language seems to make a subtle implication that for those diagnosed with AUD, the desire to use alcohol never completely goes away. Obviously, this speaks to our critique.

One additional piece of background information on the diagnosis, which is vital to our discussion, is that the above diagnosis for AUD, from the DSM-5, is a revised diagnosis. The previous version, as presented in the DSM-IV, had separate diagnoses for alcohol abuse and alcohol dependence. The DSM-5 combined both into the lone diagnosis of alcohol abuse disorder, with mild, moderate and severe severity specification (NIAAA, 2013a).

Alcoholism As a Social Construction

Social constructionism is a theoretical perspective that emerged recently, starting in the 1980's, which posits that humans collaborate to form a widely accepted understanding of their world. Gergen, an American psychologist who wrote extensively on the theory, asserted that social constructionism concerns itself with the "social processes by which theoretical and descriptive accounts come to be accepted as legitimate representations of reality (i.e., knowledge)" (as cited in Truan, 1993, p. 494).

Without getting lost in semantics, we can state more simply that according to social constructionism, our world is filled with various phenomena that exist not based on tangible, empirical evidence, but because everyone in our society, collectively, has agreed (either consciously or unconsciously) that they are so. Social constructionism is part of the post-modern school of sociology and psychology, and was developed as a way of examining the roots and effects of social phenomena such as crime, affluence or racial identity. It is frequently used as a method of reframing and reshaping contemporary issues that exploit or disempower various populations, and challenging their legitimacy. Many authors have suggested that alcoholism is an example of a social construction (May, 2001; Mulford, 1994; Reinarman, 1988; Truan, 1993).

To underscore the social constructionist perspective of alcoholism, it is important to note that our society's acceptance of the medicalization of the disorder and its treatment has occurred despite scant scientific evidence that alcohol has a demonstrable biological or genetic basis of any kind. As Suissa outlines in great detail, the only compelling scientific studies to support the existence of alcoholism as an illness are ones that show a higher incidence of alcoholism in identical twins than fraternal, and in the biological children of alcoholic parents (Suissa, 2003). Yet no study has ever found any conclusive gene for alcoholism, and despite the observed higher correlation/risk, the majority of children of alcoholics do not in fact develop an addiction to alcohol themselves. In fact, many consume alcohol without developing a problem (Peele and Brodsky, 1991).

Despite this lack of evidence, every major medical organization has endorsed the disease model of alcoholism including the World Health Organization, American Medical Association, American

Psychiatric Association and the National Association of Social Workers (Suissa, 2003). In order to understand this apparent contradiction – how our society came to such a uniform consensus on the existence of alcoholism as an illness despite a dearth of empirical evidence – we need to take a brief look at the history of alcoholism in our country.

Alcoholism in America

American society's attitude toward alcohol (and those who consume large amounts of it) evolved gradually over the last two centuries. According to Reinarman, "in the colonial era alcohol consumption was virtually universal and drunkenness common practice. Even the Puritans deemed alcohol the 'Good Creature of God.' " (1988, p. 92). It was not until the 19th century, with the rise in popularity of distilled spirits and the resulting increase in public drunkenness, that first clergy and then also doctors started to publicly demonize alcohol and advocate for temperance. It was in this early phase that alcoholism's social construction as a medical condition first emerged, including the notion that a person could have a sickness that made them powerless to resist alcohol. We also see the first documentation of the cluster of symptoms that we can still recognize today in the DSM diagnosis including cravings, tolerance, the progressive nature of alcohol consumption, loss of control, and negative social consequences (White, 2000).

The popularity of the disease model waned in the period that followed, as America went through the upheaval associated with the introduction and eventual repeal of Prohibition, as well as the Great Depression and World War II. Society still increasingly acknowledged alcoholism as a condition, but as opposed to viewing an alcoholic as a person with a sickness, during this time alcoholics were demonized as moral failures, guilty of displaying weak character and failing societal obligations. Then in the mid-20th century, the pendulum swung back and, motivated by a desire to treat people with addictions more humanely and provide them with assistance, the disease model of alcoholism rose again as the dominant view. Mulford (1994) states, "it fit well with the public's hunger for a simple explanation of public drunkenness, its yearning for simple solutions, its faith in science and technology to find a quick fix, and its awareness of medical science's dramatic successes in conquering many diseases." (p. 518).

It was also around this time that Alcoholics Anonymous (AA) was founded. The twelve-step program of AA dovetailed perfectly with the disease model of alcoholism, and its popularity soared. As virtually everyone is aware, the foundation of the twelve-step model is that the problem drinker admits and accepts complete powerlessness over his or her addiction to alcohol. Over the latter half of the 20th century, the disease model and this twelve-step approach have become by far the dominant paradigm.

Current Treatment Environment

As outlined above, in America the disease model is by far the most commonly accepted view of alcohol use disorder, and it is an understatement to say that it affects the current treatment environment. Treatment of alcoholism is now widely viewed as the responsibility of the medical profession, with a heavy reliance on twelve-step programs. In this prevailing atmosphere, it is most important to note that there is no such thing as a cure for alcoholism, only continued treatment. Once a person receives a diagnosis for AUD, he or she is typically labeled an alcoholic in perpetuity, both in his or her own mind and the minds of others, with the substantial stigma that accompanies the label. Stanton Peele, one of the foremost critics of the disease model, describes how "no matter how many years ago your Uncle Joe had his last drink, he is still considered an alcoholic. The very word addict confers an identity that admits no other possibilities. It incorporates the assumption that you can't, or won't, change." (2004, p. 43). While this may be an effective approach for older adults struggling with a lifelong addiction to alcohol (such as the creators and first members of AA), it obviously has troubling ramifications for adolescents and young adults who receive a diagnosis.

The disease model has a litany of disadvantages as it applies to adolescents and young adults, which are well captured by Peele and Broadsky in their book *The Truth About Addiction and Recovery* (1991):

First, they enter a world that tells them they have no personal control, that there is something inside of them that they can never change. It offers zero hope for the future outside of complete abstinence. This is profoundly disempowering, and can become a self-fulfilling prophecy of alcohol abuse.

Secondly, it fails to differentiate between a person who has abused alcohol for decades, and a person who has only recently developed a problematic relationship with drinking.

Third, as already discussed, it stigmatizes people in their own minds, for the rest of their life.

Fourth, it interrupts normal maturation processes, and fails to acknowledge that drinking behavior is likely to change naturally as young people enter different phases of their adult development.

Fifth, it removes youths from many typical social situations, and creates feelings of alienation, isolation and loneliness that can only make them more susceptible to problem drinking.

And finally, it limits their social interactions primarily to other recovering alcoholics and addicts who only reinforce the prominence of drinking in their lives.

It should be noted that although the disease model and twelve-step approach dominate the current treatment environment, there are alternatives. One competing model of addiction is called the adaptive model, which views alcoholism and other addictions as a functional adaptation to distress. The model suggests that an early predisposition due to genetic susceptibility, faulty upbringing and other environmental factors leads to a failure to reach appropriate levels of adult integration in society, manifested by things like economic dependence, family breakdown, self-hate, depression, or aggressiveness. Alcoholism arises as an adaptation to the stress and other negative feelings that result from their failure (Alexander, 1988). A huge advantage of a competing model such as this is that it puts problematic drinking in the context of life circumstances and emotional development, and offers hope of a permanent recovery if underlying issues are addressed. It is not nearly as fatalistic and disempowering as the disease model.

American Youth Culture and Alcohol Abuse

The prevalence of alcoholism varies widely based on factors like age, gender, ethnicity, economic status and sexuality. For example, data has long shown that alcoholism is significantly more prevalent in males than in females (APA, 2013). Alcoholism also tends to be more commonly found in younger people than older people, and in people from poorer economic backgrounds (Peele & Broadsky, 1991). Another example: several studies have shown a higher rate of alcoholism among sexual minorities, especially lesbian women (Trocki, Drabble & Midanik, 2005). Researchers have conducted thousands of studies trying to pinpoint these differences, and illuminate the causes behind the correlation.

Cultural norms with regard to alcohol, including both drinking behavior as well as attitudes toward alcohol consumption, also vary widely. This has a profound influence on a person's risk of developing alcoholism. One example cited by Peele: "research inevitably finds that Irish and Native Americans have very high alcoholism rates, and that Slavs, the English, and some American

Protestant drinkers are also at high risk for alcoholism. The Italians, Jews, and Greeks and the Chinese have exceedingly low rates of alcoholism." (1991, p. 51). He goes on to explain that this is largely due to how families expose their children to alcohol, and what children are taught culturally about its value or its danger. These cultural norms need to be taken into account when assessing and diagnosing problematic alcohol use, because they can affect the likelihood of developing a serious problem.

Currently among America's youth, there is evidence of a strong cultural focus on excessive drinking, across a range of ethnicities. This is particularly prevalent on college campuses and in depictions of youth alcohol consumption in movies and on television, where binge drinking (consuming five or more drinks in a brief period of time) and hard partying culture are norms. Here are some facts on youth drinking culture from the National Institute on Alcohol Abuse and Alcoholism:

- By age 15, more than 50% of teens have had at least 1 drink.
- By age 18, more than 70% of teens have had at least 1 drink.
- People ages 12-20 drink 11% of all alcohol consumed in the United States.
- 6.9 million young people reported binge drinking at least once in the past month.
  (NIAA, 2013b)

Given those statistics, it seems easy to imagine a young person who participates in this drinking culture who fits at least two of the DSM-5 symptoms for AUD, which would qualify them for a diagnosis of mild AUD. Indeed, in their analysis of the diagnosis and assessment of AUD among adolescents, Martin & Winters (1998) discuss how common it is for adolescent drinkers to develop some tolerance to alcohol's effects, and how low the specificity is for that criterion with the adolescent population. They note, "although marked tolerance to alcohol is an important aspect of alcohol dependence, difficulty in specifying and measure this phenomenon makes it a problematic symptom for adolescents." (p. 98).

If a college student were being assessed, and answered honestly, he or she might admit to a whole range of symptoms including drinking more than originally intended, spending large amounts of time using alcohol, failure to fulfill school or work obligations because of alcohol, giving up important activities to recover from a hangover, using alcohol in situations in which it is hazardous, or developing a tolerance (which, again, might not seem particularly difficult to imagine if you were an undergraduate student in America, or know one). All six of those are symptoms of AUD in the DSM-5, which according to the diagnostic criteria would indicate *severe* AUD.

This leads us to a very important question – is an AUD diagnosis accurate for these types of drinkers? It is certainly not an easy question to answer. They definitely exhibit signs of alcohol abuse, but considering the treatment environment for alcoholics, a diagnosis is likely to do more harm than good. I would assert that this population would be misdiagnosed as suffering from AUD. This assertion is based on a key fact: a large majority of people who report this type of problem drinking recover on their own, without formal treatment (Peele, 2004). Although society's perception may be that these youths are all on the slippery slope toward alcoholism, the numbers do not support that fear. A recently publicized study showed that only a small minority of people who consume large amounts of alcohol end up developing a problem (Esser, Hedden & Kanny, 2014). It typically requires years of problematic behavior to manifest alcohol dependence. In these instances, with adolescents and young adults, the more common outcome is to participate in the normal cultural experience when they are that age, and then mature out of it when their social setting changes and societal pressures change (full time job, raising kids, etc.). Subjecting them to formal treatment seems not only unnecessary, but given the associated stigma it seems ethically questionable.

Conclusion and Recommendations

Having proceeded through all of the above, a reader may reasonably wonder at this point if the author means to suggest that the DSM's diagnosis for Alcohol Use Disorder lacks any value at all, but that is far from the case.

While the notion of alcoholism as an illness is largely a social construction, alcohol abuse and addiction are still very real and serious problems. Many people benefit substantially from treatment, including thousands who have been successful using a twelve-step program, and treatment requires a diagnosis. Including content in the DSM to facilitate this type of diagnosis is valid and necessary. This critique is focused on the decision to combine all problematic alcohol consumption into a single disorder, and the persistent popularity of the disease model of treatment, which negatively impacts the manner in which the diagnosis is applied to adolescents and young adults.

By creating a single category for all problematic alcohol use, the DSM-5 makes alarmingly little distinction between a youth who abuses alcohol in a cultural setting where that is considered the norm, and an older alcoholic suffering from a lifetime addiction. Based on the potential negative consequences, I strongly advocate for caution, restraint and collaboration in making an AUD diagnosis. It should never be based solely on a rote progression through a checklist of symptoms. Clinicians should be fully aware of the treatment environment as well as the potential stigma and disempowerment issues, and strive to reserve a diagnosis only for clients who need it and will benefit from treatment. In the event of a diagnosis, clinicians should be willing to advocate for a variety of treatment models, including those that seek to address underlying behaviors, such as the adaptive model or harm reduction techniques. The DSM should consider revising their diagnostic criteria in a way that reduces the chances of a misdiagnosis.

Clinicians dealing with adolescents and young adults who are demonstrating problematic drinking behavior should be especially mindful of the impact of an AUD diagnosis for their clients, and should give extra consideration to the influence of youth drinking culture on their clients' behavior. Clinicians need to be mindful that recent data suggest that although there is a positive correlation between binge drinking frequency and alcoholism, only a very small minority of people who consume large amounts of alcohol end up developing a problem (Esser, Hedden & Kanny, 2014), and many youths will naturally reduce their alcohol consumption as they age, due to normal societal pressures (Peele, 2004). Considering the prevalence of the disease model of addiction, a diagnosis of alcoholism too often sends young people into a bleak, inflexible and endless treatment setting that was not designed with them in mind. Being labeled as an alcoholic stigmatizes them and robs them of the most vital recovery tool of all – the belief that they have the ability to change. By offering a seemingly objective list of symptoms, the DSM is facilitating the stigmatization and disempowerment of these adolescents and young adults.

References

Alexander, B. (1988). The disease and adaptive models of addiction: A framework evaluation. In Peele, S. (Ed), *Visions of Addiction* (p. 45-66). Lexington, MA: D.C. Heath and Company.

American Psychiatric Association. (2013). *Diagnostic and statistical manual of mental disorders* (5th ed.). Arlington, VA: American Psychiatric Publishing.

Esser, M., Hedden, S. & Kanny, D. (2014). Prevalence of alcohol dependence among adult drinkers, 2009-2011. *Preventing Chronic Disease*, 11. 1-11.

Martin, C. & Winters, K. (1998). Diagnosis and assessment of alcohol use disorders among adolescents. *Alcohol Health & Research World*, 22(2). 95-105.

May, C. (2001). Pathology, identity and the social construction of alcohol dependence. *Sociology*, 35(2), 385-401.

Mulford, H. (1994). What if alcoholism had not been invented? The dynamics of American alcohol mythology. *Addiction*, 89(5), 517-520.

National Insitute on Alcohol Abuse & Alcoholism (2013a). Alcohol use disorder: A comparison between DSM-IV & DSM-5. NIH Publication No. 13-7999. Retrieved from: http://pubs.niaaa.nih.gov/publications/dsmfactsheet/dsmfact.pdf

National Insitute on Alcohol Abuse & Alcoholism (2013b). Underage drinking. Retrieved from: http://pubs.niaaa.nih.gov/publications/UnderageDrinking/Underage_Fact.pdf

Peele, S. and Brodsky, A. (1991). *The Truth About Addiction and Recovery*. New York, NY: Simon & Schuster.

Peele, S. (2004). The surprising truth about addiction. *Psychology Today*, May/June, 43-46.

Reinarman, C. (1988). The social construction of an alcohol problem: The case of mothers against drunk drivers and social control in the 1980s. *Theory and Society*, 17(1), 91-120.

Suissa, A. (2003). Alcoholism as a disease in North America: A critical social analysis. *Journal of Addictions Nursing*, 14(4), 201-208.

Trocki, K., Drabble, L. and Midanik, L. (2005). Use of heavier drinking contexts among heterosexuals, homosexuals and bisexuals: Results from a national household probability survey. *Journal of Studies on Alcohol*, 66(1), 105-110.

Truan, F. (1993). Addiction as a social construction: A postempirical view. *Journal of Psychology*, 127(5), 489-500.

White, W. (2000). Addiction as a disease: Birth of a concept. *Counselor*, 1(1), 46-51, 73.

Antisocial Personality Disorder

Glenda J. Reed

Introduction

The purpose of this paper is to provide an alternative perspective of the Cluster B personality disorders listed in the DSM V as Antisocial Personality Disorder (ASPD). The diagnostic criteria according to DSM V are –

A. A pervasive pattern of disregard for and violation of the rights of others, occurring since age 15 years, as indicated by three (or more) of the following:

1. Failure to conform to social norms with respect to lawful behaviors, as indicated by repeatedly performing acts that are grounds for arrest.
2. Deceitfulness, as indicated by repeated lying, use of aliases, or conning others for personal profit or pleasure.
3. Impulsivity or failure to plan ahead.
4. Irritability and aggressiveness, as indicated by repeated physical fights or assaults.
5. Reckless disregard for safety of self or others.
6. Consistent irresponsibility, as indicated by repeated failure to sustain consistent work behavior or honor financial obligations.
7. Lack of remorse, as indicated by being indifferent to or rationalizing having hurt, mistreated, or stolen from another.

B. The individual is at least age 18 years.

C. There is evidence of conduct disorder with onset before age 15 years.
D. The occurrence of antisocial behavior is not exclusively during the course of schizophrenia or bipolar disorder. (DSM V, 2013, pg. 659)

ASPD is also referred to as "psychopathy, sociopathy, or dyssocial personality disorder. (DSM V, 2013, pg. 659)" The diagnosis can only be given if the individual is at least 18 years old (Criterion B), but has a history of conduct disorder symptoms before the age of 15 (Criterion C). Similar to ASPD, conduct disorder also is a pattern of disregard for the rights of others, and violation of social rules leading to similar detrimental outcomes. "The specific behaviors characteristic of conduct disorder fall into one of four categories: aggression to people and animals, destruction of property, deceitfulness or theft, or serious violation of rules" (DSM V, 2013, pg. 660).

According to the DSM V diagnosis criteria, likelihood of conduct disorder developing into ASPD increases if a child experiences and or is exposed to adverse environmental factors. "Child abuse or neglect, unstable or erratic parenting, or inconsistent parental discipline may increase the likelihood that conduct disorder will evolve into antisocial personality disorder" (DSM V, 2013, pg. 660).

The DSM V explains that individuals with ASPD tend not to conform to societal rules and often participate in unlawful behavior (Criterion A1). "They may repeatedly perform acts that are grounds for arrest (whether they are arrested or not), such as destroying property, harassing others, stealing, or pursuing illegal occupations" (DSM V, 2013, pg. 660). Persons with this disorder often participate in social interactions that make them untrustworthy. "They are frequently deceitful and manipulative in order to gain personal profit or pleasure (e.g., to obtain money, sex, or power). (Criterion A2).

They may repeatedly lie, use an alias, con others, or malinger" (DSM V, 2013, pg. 660). Individuals with ASPD also tend to be irresponsible and uncaring in their personal lives. "They may engage in sexual behavior or substance use that has a high risk for harmful consequences. They may neglect or fail to care for a child in a way that puts the child in danger" (DSM V, 2013). Individuals with ASPD also tend to be unwilling to accept accountability for the consequences of their actions and completely disregard the pain and suffering they have caused others.

> Individuals with antisocial personality disorder show little remorse for the consequences of their acts (Criterion A7). They may be indifferent to, or provide a superficial rationalization for, having hurt, mistreated, or stolen from someone (e.g., "life's unfair," "losers deserve to lose"). These individuals may blame the victims for being foolish, helpless, or deserving their fate (e.g., "he had it coming anyway"); they may minimize the harmful consequences of their actions; or they may simply indicate complete indifference. They generally fail to compensate or make amends for their behavior. They may believe that everyone is out to "help number one" and that one should stop at nothing to avoid being pushed around. (DSM V, 2013, pg. 660)

Individuals with Antisocial Personality Disorder can be deceptive and appear normal and intelligent. "They may display a glib, superficial charm and can be quite voluble and verbally facile (e.g., using technical terms or jargon that might impress someone who is unfamiliar with the topic)" (DSM V, 2013, pg. 660). Alternatively, they may also experience depression and have associated disorders such as substance abuse and anxiety. Individuals often can have features that meet the DSM criteria for other personality disorders such as borderline, histrionic, and narcissistic personality disorders. "Individuals with antisocial personality disorder may also experience dysphoria, including complaints of tension, inability to tolerate boredom, and depressed mood. Individuals with antisocial personality disorder also often have personality features that meet criteria for other personality disorders, particularly borderline, histrionic, and narcissistic personality disorders" (DSM V, 2013, pg. 660).

The DSM states that ASPD tends to be associated with lower socioeconomic status. "Antisocial personality disorder appears to be associated with low socioeconomic status and urban settings" (DSM V, 2013, pg. 662). The DSM V, using criteria from previous versions of the DSM, suggests ASPD comprises between 0.2% and 3.3% of the population. According to the DSM V, ASPD is more common in males than females. "The highest prevalence of antisocial personality disorder (greater than 70%) is among most severe samples of males with alcohol use disorder and more prevalent in lower economic socioeconomic groups" (DSM V, 2013, pg. 662). There is a genetic risk of having ASPD if the one or both biological parents also have the disorder. "The risk to biological relatives of females with the disorder tends to be higher than the risk to biological relatives of males with the disorder" (DSM V, pg. 662). The DSM V explains that biology is a factor, but environmental factors can also contribute to developing ASPD as both biological and adopted children are at an increased risk if they are exposed to childhood abuse and neglect.

> Adoption studies indicate that both genetic and environmental factors contribute to the risk of developing antisocial personality disorder. Both adopted and biological children of parents with antisocial personality disorder have an increased risk of developing antisocial personality disorder, somatic symptom disorder, and substance use disorders. (DSM V, 2013, pg. 661)

Control Theory

The premise of the DSM V criteria is that individuals who engage in behaviors listed in the DSM criteria are, in fact, controlled by that disorder. Control theory's premise is that humans are not inherently designed to conform to societal norms and rules. Humans conform to societal norms only through a series of controls, described by Hirschi below. Control theory asserts that the cause for deviant behaviors can be explained as a combination of criminal opportunity and low self-control. "The theory rests on the Hobbesian assumption that human behavior is not inherently conforming, 'but that we are all animals and thus naturally capable of committing criminal acts.'" (Wiatrowski, 1981). Control theory was developed by criminologists Travis Hirschi and Michael R. Gottfredson, who defined crime as "acts of force or fraud undertaken in pursuit of self-interest." (Encyclopedia Britannica, 2014). Control theory explains that the reason most people do not engage in the deviant behaviors is: 1) opportunity; 2) fear of punishment; 3) embarrassment; 4) lack of support; 5) learned values; 6) conscience; and 7) disapproval from loved ones (Wiatrowski, 1981).

Travis Hirschi (1963) proposed four elements of the bond to society that prevented most people from committing crimes and practicing deviant behavior:

Attachment – This is the bond we have with the closest people in our life whom we do not want to cause any pain, embarrassment or disapproval. For example, a person's mother would be disappointed if he or she were arrested for shoplifting or getting fired for stealing on the job. The disappointment and hurt we would could cause a loved one makes individuals think twice before acting impulsively.

Commitment – This is appreciating the value of something you have worked hard for and not risking throwing everything away to satisfy need(s). For example, a married man or woman would not risk having an affair because the consequences of their spouse finding out and leading to an expensive and lengthy divorce.

Involvement – Being active and keeping busy means little time to get into trouble. Boredom and inactivity are blamed for the high crime rate in poor neighborhoods as many do not have jobs or money which leads to anger, frustration and for too many crime.

Beliefs – Core values and having integrity are taught and also learned. There are some people who struggle every day financially, but would not steal a cent even if a door was left wide open so they could take it. Their beliefs and ethics will not allow them to steal or lie, but there are some people who steal even though they do not need to. These people steal because they have the power and opportunity to get away with it.

Hirschi (1963) explains that when one of these elements breaks, the temptation to lose control increases and engaging in deviant behavior can happen. For example, if a young girl is disconnected from her family because she runs away due to abuse or some other tragedy, she may then turn to prostitution, drugs and other abusive behaviors. Without these values, her beliefs come into question, as demonstrated through her changing behavior.

Anti-Social Personality Behavior: Disorder or Human Nature?

It has been my personal experience that there are many people who fit the criteria listed in the DSM V for ASPD but who do not have a mental illness. Rather, they seize the opportunity to act irresponsibly for their own personal satisfaction. They do not care who they hurt, and they will continue until they are caught and punished. Defining someone as "evil" is not a term that should be used professionally as it is judgmental, insensitive, and archaic. Merriam Webster (2014) defines evil as "arising from actual or imputed bad character or conduct." There are people who fit the definition of evil who display the same behaviors as described in the DSM V for ASPD. Would a diagnosis of ASPD describe everyone with those characteristics?

One of my arguments with the DSM V's criteria for the diagnosis for ASPD is that it does not acknowledge that people have options and can practice self-control over their actions. It is my opinion that people who consistently harm others and derive pleasure from someone's pain and suffering are evil and in most cases, not mentally ill. The DSM appears to lump any morally wrong behavior together and give the diagnosis of ASPD. In my opinion, while there are people who can be diagnosed with this disorder, ASPD should be applied to the most extreme cases such as Jeffrey Dahmer, John Wayne Gacy, Ted Bundy, etc. This is because these cases are rare and so extreme and heinous, that a person would have to be mentally ill to commit these acts of these serial killers. "Serial killers suffer from antisocial personality disorder, but then again, so do several million people in the United States. The most extreme killers fall under the sociopathic or psychopathic personality disorder classification like Ted Bundy for example" (Mayo Clinic, 2008).

There are people who are habitual liars, show total disregard for the law and safety of others, rationalize their actions by blaming others and show no remorse or guilt for the pain and suffering caused. According to DSM standards, these are some of the standards that would diagnose them with ASPD and therefore put them in a category of having a disorder. Many of these people are manipulative and clever using these skills to gain someone's trust and then taking advantage of the situation. Growing up, the saying was "if you find a fool, bump his head." This cliché was an alternative way to encourage a person to find an individual who is naïve, gain his or her trust, and then take advantage of that trust.

Thomas Szasz argued that mental illness is a myth. As an article in Psychology Today stated, "One of Szasz's basic arguments is that mental illness is a myth. He was highly critical of the so-called medical model for understanding human struggles and difficulties. He saw the uses of diagnostic systems (such as the DSM) as wrongly implying the presence of actual disease. Furthermore, he saw such efforts as medicalizing morality and the typical dilemmas and struggles of human life" (Paulson, 2012).

Szasz argued that mental illness is a myth, and that the terms sociopath and antisocial personality disorder are judgments of moral standards when a person commits acts that are evil and against social norms. Szasz explains that many evil people do not suffer from any form of psychosis, brain damage or childhood abuse and exhibit the same actions mentioned in the DSM. Szasz had little respect for the DSM and its diagnosis as it relates to mental illness:

> The primary function and goal of the DSMs is to lend credibility to the claim that certain behaviors, or more correctly, misbehaviors, are mental disorders and that such disorders are, therefore, medical diseases. Thus, pathological gambling enjoys the same status as myocardial infarction (blood clot in heart artery). In effect, the APA maintains that betting is something the patient cannot control; and that, generally, all psychiatric 'symptoms' or 'disorders' are outside the patient's control. I reject that claim as patently false. (Paulson, 2012)

Szasz argued that modern psychiatry cannot prove that the behaviors they list in the DSM cannot be self-controlled or that the behaviors are due to a chemical imbalance. Szasz believes that much diagnosis is driven by greed and powers of the pharmaceutical industry, politics and the prison system. Szasz stated that the persistent habit of some behaviors are not due to mental illness, but lack of self-control, moral character and the choices people make to satisfy their own needs (Paulson, 2012).

This writer disagrees with the assessment that ASPD is more commonly represented in lower socioeconomic groups. These behaviors occur across all economic lives as exhibited by churches, police officers, politicians and celebrities. The following examples strengthen this writer's argument that these behaviors are not always the result of mental illness, but rather a product of a person's access to power and opportunity. For example, the sexual abuse of the Catholic Church by some priests fits the behaviors in the criteria. The police officers who killed Michael Brown and Eric Garner, along with the other police brutality cases fits the criteria for ASPD also. However, these cases can also be examples of opportunity, power and evil. The police officers and clergy showed recklessness, aggressiveness and total disregard for others' safety (all DSM criteria for ASPD). Some politicians fit many of the criteria listed in DSM V, but would they be considered mentally ill, or greedy, selfish, lying opportunists that consistently take advantage of the trust and hopeless desperation of the people they promise to represent.

President Richard Nixon's actions as our president were unforgivable. He lied and manipulated an entire country for his own personal gain. Which category does Richard Nixon fall under? He exhibited many of the behaviors of the DSM, but was it corruption and greed or ASPD? The countless stories of the so called "beautiful" people of Hollywood fit the ASPD criteria. Hopefully, the Bill Cosby rape allegations are false, but one thing is certain, he has committed adultery many times. His infidelities more than likely have hurt his wife and his family, but that has not stopped him from consistently committing this behavior. He had a choice, but he also has opportunity, money and fame. Bill Cosby is still one of my childhood heroes, but he is also human, a man with flaws and weaknesses no one understands but him. It is my professional opinion that he is not mentally ill. By stating that ASPD is more commonly found in people of lower socioeconomic status, the DSM inappropriately links poverty with lack of empathy and thus with lying, aggressiveness, disregard for others, and even serial killers. The DSM ignores the poor people who would never abuse a child, take nude pictures, cheat on their significant other, lie, cheat or steal no matter the background or circumstance.

Below is a case study taken from an online article (Biography, 2014) designed to show the difference between someone with ASPD and a person who takes advantage of their power and opportunity without regard for others.

Case Study – Bernie Madoff

Bernard began to study law at Brooklyn Law School, but quit later that year to begin his own investment firm. Using the $5,000 he earned from his summer lifeguarding job and a side gig installing sprinkler systems, Madoff and his wife founded Bernard L. Madoff Investment Securities, LLC.

With the help of Madoff's father-in-law, a retired C.P.A., the business attracted investors through word-of-mouth and amassed an impressive client list including stars such as Steven Spielberg, Kevin Bacon and Kyra Sedgewick. Madoff Investment Securities grew famous for its reliable annual returns of 10 percent or more and, by the 1980s, his firm handled up to 5 percent of the trading on the New York Stock Exchange

As Madoff's fame as a successful investor grew, Madoff Securities began using computer technology to develop stock quotes. The program that the firm tested and helped to develop became the National Association of Securities Dealers Automated Quotations, or NASDAQ. Madoff later served as president of the board of directors for the NASDAQ stock exchange.

As the business expanded, Madoff began employing more and more of his family members to help with the company. His brother Peter joined him in the business in 1970 as the firm's chief compliance officer. Later, Madoff's sons Andrew and Mark also worked for the company as traders. Peter's daughter, Shana, became a rules-compliance lawyer for the trading division of her uncle's firm, and his son, Roger, joined the firm before his death in 2006.

Arrest

But Madoff became famous for a very different reason on December 10, 2008. After the investor informed his sons that he planned to give out several million dollars in bonuses two months earlier than scheduled, they demanded to know where the money was coming from. Madoff then admitted that a branch of his firm was actually an elaborate Ponzi scheme. Madoff's sons reported their father to federal authorities, and the next day Madoff was arrested and charged with securities fraud.

Madoff reportedly admitted to investigators that he had lost $50 billion of his investors' money, and pled guilty to 11 felony counts—securities fraud, investment adviser fraud, mail fraud, wire fraud, three counts of money laundering, false statements, perjury, false filings with the United States Securities and Exchange Commission (SEC), and theft from an employee benefit plan—on March 12, 2009. While the extent of his fraud is still being uncovered, prosecutors say $170 billion moved through the principal Madoff account over decades, and that before his arrest the firm's statements showed a total of $65 billion in accounts.

Madoff was imprisoned until a sentencing hearing scheduled for June 16th. He was sentenced to 150 years in prison on June 29, 2009—the maximum possible prison sentence for the 71-year-old defendant (Biography, 2014).

Bernie Madoff displayed the behaviors listed in the DSM V. Note that he is not from a low socioeconomic group. He did hurt many people, stole their life savings, hurt his family (one of his sons committed suicide, Long, 2010). He committed these behaviors because he had the opportunity, and he made the choice to use his intelligence to cheat people who trusted and depended on him. Mr. Madoff fits the criteria of the DSM V, but should he be considered mentally ill or evil? This would be a question that should be answered by the people who lost their life savings and after believing they can retire and enjoy their senior years have to return to work to survive.

According to Gabbard's *Psychodynamic Psychiatry*, antisocial personality disorder is one of the most extensively studied personality disorders. Gabbard states, "Antisocial patients are perhaps the most extensively studied of all of those with personality disorders, but they are also the patients that clinicians tend to avoid the most" (pg. 491). Based on my personal and professional experiences with individuals who habitually lie, cheat, steal, manipulate for personal gain and derive satisfaction someone else's pain, it is a challenge to trust and practice objectively. Standardized assessments may be helpful in allowing clinicians to maintain some of that objectivity. As the DSM states, "because deceit and manipulation are central features of antisocial personality disorder, it may be especially helpful to integrate information acquired from systematic clinical assessment with information collected from collateral sources" (DSM V, 2013, pg. 662).

Conclusion

It is this writer's opinion that it is not a sickness that causes somebody to engage in aberrant behavior. Whether an individual has a serotonin rush in the brain or he needs a certain kind of adrenaline flow in order to be happy, a person makes choices. The DSM V has more work to do in explaining the diagnoses and behaviors. There are millions of people who are abused, neglected and poor who are still committed to society through Hirschi's elements. These people do not intentionally plan to hurt another person or thing – especially if that person or thing is less vulnerable even though their path has been incredibly difficult. The DSM V does not give humans the credit of knowing the difference between right and wrong and then making an ethical judgment when they are faced with a choice. This writer believes that we all have the propensity to hurt someone else when we are angry, cheat on our significant other if we know we will not get caught, or steal from the cash drawer at work when no one is looking and we have no money. But for most people we make the choice not to because we do not want to get caught, embarrass our loved ones, face punishment and our ethical standards. There are some that take risks that will hurt many, but they could not care less. In many circumstances, it is petty, but dishonest. Then there are other cases, such as Bernie Madoff, some police officers, and clergy where it is more serious and hurtful. They are taking what they want and putting it above what society says. This is evil behavior that should not be excused by the diagnosis of a mental illness.

References

American Psychiatric Association. (2013). *Diagnostic and statistical manual of mental disorders: DSM-5*. Washington, D.C: American Psychiatric Association.

Bernard Lawrence Madoff. (2014). *The Biography.com* website. Retrieved 09:30, Dec 05, 2014, from http://www.biography.com/people/bernard-madoff-466366Gabbard, G. (2000). *Psychodynamic psychiatry*, Third Edition. Washington, D.C: American Psychiatric Press, Inc.

Long, C. (2010). Mark Madoff Suicide: Bernie Madoff son found hanged in NYC apartment. Huffington Post. http://www.huffingtonpost.com/2010/12/11/mark-madoff-suicide-hanged_n_795342.html

Mayo Clinic Staff. (2008, October 8). *Antisocial personality disorder.* Retrieved from http://www.mayoclinic.com/health/antisocial-personality-disorder/DS00829

Merriam Webster. (2014). http://www.merriam-webster.com/dictionary/evil

Poulsen, B. (2012). *Revisiting the myth of mental illness: some thoughts on Thomas Szasz, Psychology Today* Retrieved from http://www.psychologytoday.com/blog/reality-play/201209/revisiting-the-myth-mental-illness-some-thoughts-thomas-szasz

Travis Hirschi. (2014). In *Encyclopedia Britannica.* Retrieved from http://www.britannica.com/EBchecked/topic/1340901/Travis-Hirschi

Wiatrowski, M. et al. (1981). Social control theory and delinquency. *American Sociological Review*, Vol. 46, No. 5 (Oct., 1981), pp. 525-541. Published by: American Sociological Association. Article Stable URL: http://www.jstor.org/stable/2094936

Antisocial Personality Disorder and Attachment Theory.

Kelly Trentz

When thinking about labeling a person with a diagnosis there are certain criteria that a person must fulfill in order to have that disorder. The professionals who are diagnosing people with mental illness are clustering together the client's symptoms to fit into a category that is a diagnosis. When reviewing the criteria for some disorders, it seems that the criteria can be very particular but can also be very vague at the same time. When reviewing the criteria for personality disorders in the DSM 5, I noticed that the criteria were very strict on what a person's symptoms are in order to have a diagnosis, but there is no mention of any external factors that can be contributing to an impairment that the person is displaying at that time.

The DSM 5, also known as the *Diagnostic And Statistical Manual Of Mental Disorders*, contains diagnostic criteria for all diagnosed mental illnesses. This manual is used to diagnose as well as to understand symptoms of a diagnosis when working with mental health. Incorporated in the manual is a section for personality disorders, and the criteria for each specific personality disorder. According to this manual, a personality disorder is an enduring pattern or inner experience or behavior that deviates markedly from an individual's culture, is pervasive and inflexible, has an onset in early adulthood, is stable over time and leads to distress or impairment. One of the more specific personality disorders that are referenced in the DSM 5 is Antisocial Personality disorder. Encompassed with the diagnosis of Antisocial Personality Disorder is the diagnostic criteria which a person may display to be diagnosed with that specific disorder. The criteria for this disorder are as follows:

    A.) Moderate or greater impairment in personality functioning, manifested by characteristic difficulties in two or more of the following four areas:

        1.) Identity: Egocentric; self-esteem derived from personal gain, power, or pleasure.

        2.) Self-Direction: Goal setting based on personal gratification; absence of prosocial internal standards, associated with failure to conform to lawful or culturally normative ethical behavior.

        3.) Empathy: Lack of concern for feelings, needs, or suffering of others; lack of remorse after hurting or mistreating another.

        4.) Intimacy: Incapacity for mutually intimate relationships, as exploitation is a primary means to of relating to others, including by deceit or coercion, use of dominance or intimidation to control others.

    B.) Six or more of the following seven pathological personality traits:

        1.) Manipulativeness: Frequent use of subterfuge to influence or control others; use of seduction; charm; glibness; or ingratiation to achieve one's ends.

        2.) Callousness: Lack of concern for feelings or problems of others; lack of guilt or remorse about the negative or harmful effects of ones actions on others; aggression; sadism.

3.) Deceitfulness: Dishonesty and fraudulence; misrepresentation of self, embellishment or fabrication when revealing events.

4.) Hostility: Persistent or frequent angry feelings; anger or irritability in response to minor slights and insult; mean, nasty, or vengeful behavior.

5.) Risk Taking: Engagement in dangerous, risky, and potentially self-damaging activities, unnecessarily and without regard for consequences; boredom proneness and thoughtless initiation of activities to counter boredom; lack of concern for one's limitations and denial of reality or personal danger.

6.) Impulsivity: Acting on the spur of the moment in the response to immediate stimuli; acting on a momentary basis without a plan or consideration of outcomes; difficulty establishing and following plans.

7.) Irresponsibility: Disregard for- and failure to honor financial and other obligations or commitments; lack of respect for – and lack of following through on- agreements and promises.

C.) Evidence of Conduct Disorder with onset before age 15 years old.

Note: Individual must be at least 18 years of age.

These diagnostic criteria are the outline for diagnosing a person with antisocial personality disorder. A person must display two components in section A, and six components in section B. Although it is very specific on what is essential for the diagnosis to be recognized as authentic, I believe that there are some controversial points displayed within the criteria in addition to not incorporating external factors that may influence a persons' personality. In Section B of the criteria there are a total of seven factors that constitute pathological personality traits, and in order to be considered for a diagnosis a person must present with six of these traits.

When reviewing the components that make up section B, these are normally not areas of a person's personality that are at the forefront when performing an assessment. Depending on the individual and how honest they are about their interactions with others, usually a person would not recognize these traits early on in a professional/clinical relationship. Discussing the client's symptoms with people who interact with the client regularly will help to uncover these types of personality traits, but also working with the client for a longer period of time will help to uncover a proper diagnosis of antisocial personality disorder.

According to the DSM 5, sometimes when people have a diagnosis at a younger age of conduct disorder, that diagnosis may progress into antisocial personality disorder. This is not the case for every child who receives a diagnosis of conduct disorder, but it can be helpful during an assessment to have information regarding past diagnosis if one exists. I believe that as part of the diagnosis the DSM 5 should have elaborated in the criteria that previous evidence of a diagnosis of a conduct disorder can anticipate that antisocial personality disorder may be present, but not every person diagnosed with conduct disorder will have antisocial personality disorder in adulthood.

The diagnostic criteria for conduct disorder can look similar to antisocial personality disorder in the regard that the criteria contain a disregard for the basic rights of others, lack of remorse, as well as incorporating deceitfulness and violation of social norms. When thinking about a person's personality and the traits that they develop, it brings up some questions regarding how these traits were acquired and where the person learned certain behaviors.

When trying to understand antisocial personality disorder, I have attempted to apply a theory known as attachment theory to the diagnosis. First, it is important to understand the definition of attachment; attachment according to John Bowlby, in the relational context is an emotional bond with another person. This theory is centered on the relationship between people, specifically looking at early relationships with parents and/or caregivers. John Bowlby was a psychiatrist who helped introduce attachment theory. According to Bowlby's, *A Secure Base: Parent-Child Attachment And Healthy Human Development*, he believed that attachment theory has three main emphasises, which are:

a.) The primary status and biological function of intimate emotional bonds between individuals, the making and maintaining of which are populated to be controlled by a cybernetic system situated within the central nervous system, utilizing working models of self and attachment figure in relationship with each other.

b.) The powerful knowledge of infant and child development of the ways he is treated by his parents, especially his mother figure.

c.) The present knowledge of infant and child development requires that a theory of development pathways should replace theories that invoke specific phases of development in which it is held a person may become fixated and./or to which he may regress.

These points help to explain the root of attachment theory and how it can be applied to a person's developmental stages. When conducting research regarding this theory Bowlby looked at early relationships between children and primarily their mothers. He concluded "the infant and young child should experience a warm, intimate, and continuous relationship with his mother (or permanent mother substitute) in which both find satisfaction and enjoyment" (Bowlby, 1951, p. 13). In order for a child to have a stable emotional upbringing the child must have a reciprocated relationship of affection in early years with caregivers.

When thinking about attachment in the aspect of a relationship, there are two main categories which attachment can fall into, secure attachment and insecure attachment. According to Bowlby the characteristics of insecure attachment include: the child being overly clingy, behaviors that are rejecting by the caregiver, as well as disorganized relationships. When a caregiver displays an inconsistent reaction to a child and his or her needs it causes the child to feel insecure with the relationship, which allows for the foundation of an insecure and inadequate attachment. When the relationship is inconsistent, it prevents the child from trusting the compassion being displayed is genuine; therefore the child does not learn how to successfully regulate his or her emotions or to meet his or her own emotional needs. These inadequacies in early relationships can cause impairments in recognized personality traits included in the diagnostic criteria of the DSM 5 for antisocial personality disorder.

The other type of attachment is a secure attachment, which is characterized by creating a trusting and safe relationship for children to feel secure with caregivers (Cherry, K. 2006) . When a child feels that his or her needs are being met and that the compassion is genuine it allows him or her to regulate emotions as well as establish healthy relationships. According to Cherry, "this helps children to develop a better self-esteem, have lower levels of anxiety and depression, and have successful social relationships."

Understanding the relationships that children establish early in life can help to understand the way relationships are formed throughout a person's life. Mary Ainsworth also studied attachment theory. She proposed that there are three types of attachment styles, and identified a result based on if the attachment was secure or insecure for the child.

The stages of attachment described by Peggy Emerson and Rudolf Schaffer show the stages of children's needs as well as establishing relationships with caregivers. According to Emerson and Schaffer, there are four stages of attachment: pre-attachment (0-3 months), indiscriminate attachment (6 weeks to 7 months), discriminate attachment (7-11 months), and multiple attachment (after 9 months).

These are the proposed stages of attachment that children go through when establishing an attachment. The stages describe the phases of how a child would display a secure attachment with their caregivers. Although these stages display a healthy secure attachment, there are external factors that would affect these stages. First there needs to be an opportunity for the child to have a secure attachment while external factors come into play.

One external factor that would affect the stages of attachment is the family dynamics that are introduced to a child. A child can be a part of many different types of family structures and some of these dynamics can have a negative affect on attachment. For example, adoption is a family dynamic that can interrupt a secure attachment. The age of the child upon entering a new home can create a barrier with regards to creating a concrete secure attachment. However, adoption alone does not mean that children cannot create secure attachments, but it may create impairment in the attachment.

Another example of a family dynamic would be when a child comes from an abusive or neglectful home. In this situation, when the child does not feel safe and secure due to the environment, attachment could be impaired as well. Witnessing abuse in the home or being a victim of abuse or neglect is a message to the child that they are not safe, and that their emotional needs are not being met, which are the criteria for an insecure attachment. In addition, children who are raised in a foster home or in an orphanage, where there is not a primary caregiver, are also at risk for insecure attachments to caregivers. Factors that may affect the family dynamic differ based on the family's cultural beliefs, race, religion, class and sexual orientation. Some roles that are taken on by caregivers in the home can originate from these factors.

When reviewing the diagnostic criteria for antisocial personality disorder and the impairments within the personality, it is important to consider external factors to better understand how the client relates to the world and functions in society. By applying attachment theory to antisocial personality disorder I believe that it humanizes the clients and shows the relationship between experience and impairment. Creating a secure attachment at a young age can have many effects on a person, primarily the way relationships are created throughout his or her life. When looking at the impairments of the personality as the criteria for antisocial personality disorder, it is evident that the traits can be derived from a disrupted early emotional connection and inability to regulate emotions.

Lack of concern for others, as well as self-esteem attained by personal self-gain, can be viewed as a way of thinking that you have to look out for yourself. In that aspect, I believe that creating that state of mind that you have to look out for yourself and not feeling safe in the world stems from childhood experiences and insecure early attachment. The DSM 5 was created to be applied to a person's displayed symptoms, so I believe that the criteria for a disorder should include a more personable approach and include external factors and personal experiences that would cause a person to display certain symptoms.

References

Ainsworth, M. D. S., & Bell, S. M. (1970). Attachment, exploration, and separation: Illustrated by the behavior of one-year-olds in a strange situation. *Child Development, 41*, 49-67.

American Psychiatric Association. (2013). *Diagnostic and statistical manual of mental disorders* (5th ed.). Washington, DC:

Bowlby, J. (1988). *A secure base: Parent-child attachment and healthy human development.* New York: Basic Books.

Bowlby, J. (1951). Maternal care and mental health. *World Health Organization* `*Monograph*(Serial No. 2).

Cherry, K. A. (2006). What is attachment theory? Retrieved from http://psychology.about.com/od/loveandattraction/a/attachment01.htm

Main, M., & Solomon, J. (1990). "Procedures for identifying infants as disorganized/disoriented during the Ainsworth Strange Situation". M.T. Greenberg, D. Cicchetti & E.M. Cummings (Eds.), *Attachment in the Preschool Years* (pp. 1 21–160). Chicago, University of Chicago Press.

Schaffer, H. R. & Emerson, P. E. (1964). The development of social attachments in infancy. *Monographs of the Society for Research in Child Development, 29*, 94.

The Intersection of Borderline Personality Disorder and Social Construction

Breeanna Horton

Borderline personality disorder (BPD) exists in about 1.4% of the population. When narrowed down, that is about 1 in every 150 people (Salters-Pedneault, 2014). An incredible number of psychologists, social workers, researchers, and advocates of clients who are diagnosed with BPD are continuously seeking a direct correlation to what is causing our clients so much distress. Pointing a finger at any one concept or gene would be inadequate and would not give a person a chance to become anything but their stigmatic disorder. While people are anxiously awaiting answers about this upsetting illness, we should take a look at how people with BPD are being looked at in society. This paper will serve as a critique of the DSM's classification criteria used to describe people with BPD, as well as the lasting effects on the families and individuals.

What does classification mean? According to Sadock (2014), "Classification is the process by which the complexity of phenomena is reduced by arranging them into categories according to some established criteria for one or more purposes (p.1234)." The system of classification posed by the DSM-5 serves many purposes including providing a common source for mental health professions to use as a reference.

The DSM is a well-known manual in the helping professions that looks at the behaviors of people who are believed to have some sort of mental irregularity. Once multiple people display similar characteristics, psychologists and many others will categorize these people and categorize them by labels; for example, Borderline

Personality Disorder. If a person is diagnosed with BPD, what does it look like? Well, according to the new DSM-5, BPD has many possible characteristics. BPD can be traced to early adulthood. If five out of the nine criterions exists within a person's behavior, they are said to have Borderline Personality Disorder. The first criterion is "frantic efforts to avoid real or imagined abandonment." The second criterion is a "pattern of unstable and intense interpersonal relationships characterized by alternating between extremes of idealization and devaluation." The third criterion is an, "identity disturbance (markedly and persistently unstable self-image or sense of self)." The fourth is the "impulsivity in at least two areas that are potentially self-damaging (spending, sex, substance abuse, reckless driving, binge eating)." The fifth criterion is, "Recurrent suicidal behavior, gestures, or threats, or self-mutilating behavior." The sixth criterion includes, "Affective instability due to a marked reactivity of mood (intense dysphoria, irritability, or anxiety)." The seventh is "chronic feelings of emptiness." The eighth criterion is "Inappropriate, intense anger or difficulty controlling anger." The final criterion possible is "Transient, stress-related paranoid ideation or severe dissociative symptoms." (American Psychiatric Association, 2013). This diagnosis is intense and intimidating.

After reviewing the criteria of BPD, I find myself feeling uneasy about treating future clients who are said to have this diagnosis. The DSM-5's criterions of BPD is scarred by the unnerving and uncomfortable phrases used to describe this disorder. Terms such as frantic, unstable, paranoid, and threats stick out like a sore thumb while reading this diagnosis. To make sense of this uneasy feeling, I sought out a theory that made the most sense. The theory I chose is Social Construction Theory.

Social Constructivism is defined by Berger & Luckmann (2003) as, "a school of thought pertaining to the ways social phenomena are created, institutionalized, and made into tradition by humans." When the theory is applied to reality, it suggests that the way people present themselves to others is shaped partly by our interaction with others, as well as by our life experiences. The way in which our caretakers raised us, the beliefs they instilled in us, how we perceive others, and how others see us is all a reflection of what was socially acceptable within the home where we were raised. This theory of cognitive development rests the responsibility of learning on more than just the individual's cognitive process, but on the interaction between the individual and his or her environment. "Social constructivism can be described as socialization, a process of acquisition of skills, knowledge, and dispositions that enables the individual to participate in his or her group or society" (Sivan, 1986). When looking at mental illness as a whole, people are generally unaware of the many disorders and are unable to fully comprehend the complexity of each disorder. We latch on to media, social networking, and the culture we live in to help us define what we do not understand.

In order to make sense of this theory, I was forced to apply it to my own life experiences to have a reference point from which to draw. A few years back, during my undergraduate studies, my boyfriend (at the time), "Alex", had started displaying some odd behavior. At first, he stopped caring about his personal hygiene (which was abnormal for him). Then he stopped attending classes and began missing important exams. Alex's friends also noticed he had become socially withdrawn and had not left his room for more than five minutes at a time. One day, as his roommate went to talk with him to see if he could help, he found Alex arguing with the foosball table. No one had been around Alex, and he seemed visibly upset. His roommate later called me and explained the situation, and we both agreed we needed to talk with Alex's parents to get Alex some help. Since I was studying social work and had been doing research in the DSM-IV-TR (an older version of the DSM-5), I began searching for the answers before calling his parents. What I found was horrifying.

Given his symptoms, Schizophrenia was the most logical diagnosis provided by the DSM and online research. Once I called his parents, my learning process truly began. We took him to the emergency room, where they held him for a few hours before releasing him to a mental health hospital. Once he was there, the psychologists required him to stay for a seventy-two hour observation period while they tried to find the correct medication for his condition. The psychologists explained that they were required to hold him because he was a danger to himself and/or to others. After he was released, I became overly cautious and nervous around him. I even found myself worrying about the way our friends and family reacted around him too. After an extensive amount of research and training, I discovered that I was caught up in this social construct of mental illness. The DSM had given us a label to work with rather than a person. We examined every word and action of his as if it were a symptom of his schizophrenia instead of remembering the quirky and fun individual that Alex was and still is.

While I am aware that there are many differences between a diagnosis of BPD and Schizophrenia, the effects explained by the Social Constructivism theory can produce the same results regardless of diagnosis. When people seek answers to their questions about BPD, they often correlate the diagnosis with the mass shootings in the news, actors from movies, or a negative personal experience with someone who had been diagnosed. In social work in specific, we study the DSM-5 and work to understand a person who is diagnosed with BPD. However, the DSM-5's criterions are to be used as a reference and as a small piece of information on how to help an individual struggling with BPD. We do not want to put individuals in boxes and leave them no choice but to become anything else but their disorder. The criterions of BPD describe a person who may have an unstable interpersonal relationship, inappropriate or intense anger, chronic feelings of emptiness, frantic efforts to avoid real or imagined abandonment, and impulsivity (i.e. substance abuse, binge eating, etc.). As humans, we go through tough relationships, death, substance abuse, eating disorders, abandonment issues, etc. It is a natural human response to feel anger, sadness, and the rest of the emotional spectrum. Is it necessary to categorize the feelings we have been given as a result and turn them into a disorder?

Not only is it possible to force a diagnosis of BPD on someone who does not fully comprehend their symptoms or condition, but we might even be preventing them from seeking treatment. As a society, we are not the only ones stigmatizing individuals with Borderline Personality Disorder. Clinicians have been known to report individuals with BPD as, "not sick, manipulative, more difficult, angry, noncompliant, and hateful. (Nehls, 1998)." These terms used by clinicians show a lack of empathy and respect and prove to become an unfortunate and misguided beginning to treatment.

A study done by Gallop, Lancee, and Garfinkel found the label of BPD to impact the treatment by care providers. There were patients with BPD and schizophrenia in this specific study. The study was aimed at comparing the difference of treatment from one disorder to the other. Results indicated nurses were more likely to remain empathic and sympathetic towards patients with schizophrenia and, "made belittling or contradicting responses to statements made by patients with BPD" (Aviram, Brodsky, & Stanley, 2006). Most importantly, Gallop and his colleagues found the nurses felt they could respond in this manner due to the thought that it was acceptable to undervalue patients with BPD (Aviram, Brodsky, & Stanley, 2006). If we are expecting individuals who are diagnosed with BPD to lash out with intense anger or wait for signs of suicidal behavior, are we really giving the individual a chance to become a well-adjusted citizen? The short answer is no; we are constructing social boundaries, forcing them to be a diagnosis and not a real human who makes mistakes.

Practitioners are not the only ones who could be giving clients with BPD the cold shoulder. Families and friends may also look the other way if an individual is diagnosed. They might do some research in the DSM or look online to find out more information about Borderline Personality Disorder. As I did my own research, I put BPD in the search engine to see what would pop up. In the images category, a lot of angry and troubled faces were displayed side-by-side with phrases such as: life of the party, distressed, manipulative, drama queen, etc. In regards to the web feedback from the search engine results, a lot of mental health websites showed up, so it is possible for people to find out more about the disorder. The quick search I did proved to me that BPD has a lot of media-provided information. But, how much of it is working against the stigma associated with this illness?

One website I came across was called Project Borderline. The website is connected to a foundation led by NFL player Brandon Marshall. Brandon was diagnosed with BPD and came forward earlier this year to raise awareness. He launched this website in order to spread the word, fight stigma, educate, advocate, reach out, bridge the gap, and change the face and future of this disorder. Years back, I remember watching celebrity gossip television programs and Brandon Marshall was something of a regular. He had been in trouble with the law for domestic disputes, drinking while driving, and multiple physical altercations. Seeing the positive change (via media) that Brandon Marshall has made over the years, on top of becoming a very outspoken advocate, he has proven that it is possible to not only overcome the symptoms of the disorder, but also break free of the stigma associated with BPD. He has spoken publicly about his illness and has helped raise awareness through his foundation and by supporting many events hosted by the National Alliance of Mental Illness on behalf of those diagnosed with a mental illness. However, consideration must to be given to the notion that having Brandon

Marshall as a major spokesperson for BPD may not be a stride toward moving away from stigma. Because he was in the media for his violent outbursts and inappropriate behavior, people might use him as an example of how people with BPD are predicted to behave. Should people diagnosed with BPD be judged upon the actions of those branded with the same label before them? It is also very important to note that Brandon Marshall is an outlier within the diagnosis of Borderline Personality Disorder and in the community of those struggling with mental illness.

There are many other celebrities who the media claims to have borderline personality disorder. A web search will return names like Lindsay Lohan, Marilyn Monroe, and Courtney Love. Whether or not these claims are true is a matter that could only be answered by their mental health professionals. Those four individuals are also outliers and do not predict, nor should they be an example of, the thousands of others who have and will continue to be diagnosed with BPD. More examples of outliers within the BPD population are the cases that involved mass murders. Adam Lanza and James Holmes are just two examples of terrifying cases of mass murder that were associated with BPD as a result of hype and hysteria. The media never released clear evidence of either being diagnosed with BPD. However, as soon as the crimes were committed, psychologists and other mental health professionals were consulted regarding the likelihood of the presence of Borderline Personality Disorder in either individual.

In my professional opinion, we need to stop categorizing individuals with BPD. Take a look at the individual before their diagnosis to see what character traits were present. Some might say they saw a pretty normal person who made some irrational decisions at times. Some might say they saw someone with an addiction problem, such as excessive drinking or substance abuse. We, as humans, are continuously making irrational and misguided decisions in life. Whether it is with our jobs, friends, dating, or something as small as eating more food than necessary, the potential for poor decisions is constantly present. Now, consider what changed following the diagnosis. The individual is still the same person they were before, but now there is a wealth of DSM-5 classifications trailing closely behind them. Then we attach symbols, phrases, and socially constructed ideas. These individuals are humans and deserve to be treated as such. Anyone stuck with the label of BPD is judged for every word, action, and decision in an unfair and, often, prejudicial manner that makes hasty generalizations about the person behind the condition.

In addition, it is important to consider the term surplus stigma. Issues that promote stigma and, thus, further the BPD misunderstanding include: theories on the development of the disorder, a preconceived notion that the mother or father may have a mental illness, frequent refusal by mental health professionals to treat BPD patients, negative and sometimes derogatory web site information that projects hopelessness, and clinical controversies as to whether the diagnosis is a legitimate one. This final controversy often results in a denial of reimbursement consideration by insurance companies for patients with Borderline Personality Disorder (Hoffmann, 2007).

Could the DSM-5 categories be helpful? There are some instances when the DSM-5 diagnostic criteria are advantageous; but, there are more pitfalls than benefits. It is useful when an official diagnosis is necessary for an insurance company to help pay for ongoing therapy. In many cases, if there is no diagnosis the insurance company will turn down coverage for treatment. This can then lead to a client walking away from necessary therapy. Also, some individuals need a label to lean on to understand the life changes that may occur. It is definitely possible that a person faced with the symptoms of BPD would feel much more irrational and erratic without knowing more specific details associated with their condition. However, it is flawed logic that a book could truly predict the intimate details of each individual's personal struggle with BPD. Each client should be given the chance to define themselves by their own thoughts and actions, and not by the stigma and standards set forth by any book, even the DSM-5.

References

American Psychiatric Association. (2013). *Diagnostic and Statistical Manual of Mental Disorders* (Vol. 5). Arlington, VA: American Psychiatric Association.

Aviram, R., Brodsky, B., & Stanley, B. (2006). Borderline Personality Disorder, Stigma, and Treatment Implications. *Harvard Review* , 249-256.

Berger, P., & Luckmann, T. (2003). *Social Construction.* Retrieved December 4, 2014, from Dictionary: dictionary.reference.com/browse/social+constructionism

Cobb, N. H. (2008). Cognitive-Behavioral Theory and Treatment. In N. Coady, & P. Lehmann (Eds.), *Theoretical Perspectives for Direct Social Work Practice: A Generalist-Eclectic Approach* (Vol. 2). New York, NY: Springer Publishing Company LLC.

Dolan, E. (2013, August 25). *Study Finds Wealth Gives Rise to a Sense of Entitlement and Narcissistic Behaviors.* Retrieved from www.informationclearinghouse.info/article35966.htm

Gabbard, G. O. (2005). *Psychodynamic Psychiatry in Clinical Practice* (Vol. 4). Arlington, VA: American Psychiatric Publishing Inc.

Halpenny, A., & Pettersen, J. (2014). *Introducing Piaget: A guide for practitioners and students in early years education.* New York, NY: Routledge.

Hoffmann, P. (2007). *Borderline Personality Disorder: A Most Misunderstood Illness.* Retrieved from NAMI-National Alliance of Mental Illness: www.nami.org/template.cfm?section=2007%template-/contentmanagement/contentdisplay.cfm&contentID=44745

Kashdan, T. B., Zvolensky, M. J., & McLeish, A. C. (2008). Anxiety Sensitivity and Affect Regulatory Strategies: Individual and Interactive Risk Factors for Anxiety-Related Symptoms. *Journal of Anxiety Disorders , 22* (3), 429-440.

Nehls, N. (1998). Borderline Personality Disorder: Gender Stereotypes, Stigma, and Limited System of Care. *Issues in Mental Health* , 97-112.

Robinson, O. J., Vytal, K., Cornwell, B. R., & Grillon, C. (2013, May). The impact of anxiety upon cognition: perspectives from human threat of shock studies. *Frontiers in Human Neuroscience* .

Sadock, B. *Synopsis of Psychiatry.* New York: Lippincott Williams & Wilkins.

Sivan, E. (1986). Motivation in Social Constructivist Theory. *Educational Psychologists* , 209-231.

Westra, H. A., Arkowitz, H., & Dozois, D. J. (2009). Adding a motivational interviewing pretreatment to cognitive behavioral therapy for generalized anxiety disorder: A preliminary randomized controlled trial. *Journal of Anxiety Disorders , 23* (8), 1106–1117.

Narcissistic Personality Disorder:
The Perfect Group

Quanesha Griffin

We are living in a society of privileged wealth and poverty within different ethnicities here in America. Wealth in society is a symbol of success and therefore makes an impact on narcissistic traits. This paper presents a study regarding wealth and narcissism. Piff conducted five experiments to investigate the associations between social class, entitlement, and narcissism. His theory is that privilege means upper-class people are "often left with feelings of entitlement and narcissistic tendencies" which reflects a growing body of social science research on narcissistic behavior. He suggests that social class has a powerful impact on our personalities. Also the disadvantaged are divided in some societies and come to depend on mutual relationships that can provide aid. They also become more interdependent. Upper-class individuals tend to have more "control over their life style" and a "reduced exposure to external influences," a set of life experiences that promotes a "greater independence" from others and more of a "self-focus." (Dolan, 2013)

The first experiment consisted of a survey that measured levels of entitlement and socioeconomic status. Piff found higher social class was associated with an increased sense of entitlement. Upper-class individuals were more likely to believe they deserved special treatment and feel entitled to "more of everything." They were also more likely to believe that if they were on the Titanic, they would deserve to be on the first lifeboat. In the second and third experiments, Piff used other surveys with different measures of entitlement and socioeconomic status to confirm his initial findings. In the fourth experiment, Piff discovered that upper-class individuals were more likely to look at their own reflections in a mirror, even when controlling for self-consciousness. The final experiment found that exposing upper-class individuals to egalitarian values reduced entitlement and decreased narcissism. (Dolan, 2013)

What does it mean to be born in a privileged ethnicity? I believe we live in a Narcissistic society. I believe that when individuals are privileged they may already have a preconceived belief that they deserve or have already earned their status.

"Social Darwinism, the theory that persons, groups, and races are subject to the same laws of natural selection as Charles Darwin had perceived in plants and animals in nature. The concept of social Darwinism has been said to rationalize different concepts of evolution, racism, and imperialism. The biggest negative connotation for this theory is that people consider it a rejection of compassion and social responsibility. Social Darwinists typically deny that they advocate a 'law of the jungle.' But most propose arguments that justify imbalances of power between individuals, races, and nations because they consider some people more fit to survive than others."(Bannister, 2000)

Social Darwinists hold that the life of humans in society is a struggle for existence ruled by "survival of the fittest" (Britannica, 2014). Some individuals believe that they are successful and they are the most intelligent because they happened to be born in a privileged group. I believe that the perspective of an individual being intelligent only based off of their background and ethnicity is not accurate, which causes me to further investigate Narcissistic Disorder in the Diagnostic and Statistical Manuel of Mental Disorders (DSM). By looking at the DSM 5 criteria and the Darwinian social theory together, we can see how this might affect individuals in a cultural manner. I will be exploring the correlation between Narcissistic disorder according to the DSM 5, Darwinians theory, culture, and gender.

According to the DSM 5, Narcissistic Personality Disorder is a pervasive pattern of grandiosity and lack of empathy, beginning in early adulthood and present in a variety of contexts (Association, American Psychiatric, 2013). One would be preoccupied with fantasies of unlimited success, power, brilliance, beauty, or ideal love (Association, American Psychiatric, 2013). The individual believes that he or she is "special" and unique and can only be understood by or should associate with, other special or high-status people or institutions (Association, American Psychiatric, 2013). And of course one would believe they have a sense of entitlement.

"Individuals with narcissistic personality disorder generally have a lack of empathy and have difficulty recognizing the desires, subjective experiences, and feelings of others. They may assume others are totally concerned about their welfare. They tend to discuss their own concerns in inappropriate and lengthy detail, while failing to recognize that others also have feelings and needs. These individuals are often envious of others or believe that others are envious of them. They may begrudge others their successes or possessions, feeling that they better deserve those achievements, admiration, or privileges." (Association, American Psychiatric, 2013)

When looking at the diagnostic features. individuals with this disorder frequently feel less modest about their capabilities, but also tend to inflate their accomplishments. Narcissistic individuals believe that they have seniority, are superior and more popular than the regular individuals in society. Also according to the DSM 5, the prevalence for narcissistic personality disorder ranges from 0% to 6.2% in different community samples (Association, American Psychiatric, 2013).

After reading the DSM5 criteria, I found an article suggesting shortcomings of the diagnosis for narcissistic personality disorder by A&M University.

"The criteria for personality disorders in Section II of DSM-5 have not changed from those in DSM-IV. Therefore, the diagnosis of Section II narcissistic personality disorder (NPD) will perpetuate all of the well-enumerated shortcomings associated with the diagnosis since DSM-III. The news reporters obtained a quote from the research from Texas A&M University, 'In this article, we will briefly review problems associated with Section II NPD and then discuss the evolution of a new model of personality disorder and the place in the model of pathological narcissism and NPD. The new model was intended to be the official approach to the diagnosis of personality pathology in DSM-5, but was ultimately placed as an alternative in Section III for further study. The new model is a categorical-dimensional hybrid based on the assessment of core elements of personality functioning and of pathological personality traits. The specific criteria for NPD were intended to rectify some of the shortcomings of the DSM-IV representation by acknowledging both grandiose and vulnerable aspects, overt and covert presentations, and the dimensionality of narcissism. In addition, criteria were assigned and diagnostic thresholds set based on empirical data.' According to the news reporters, the research concluded: 'The Section III representation of narcissistic phenomena using dimensions of self and interpersonal functioning and relevant traits offers a significant improvement over Section II NPD' " (Bender, 2014).

Looking at the gender of those diagnosed with Narcissistic personality disorder 50% to 75% are male (Association, American Psychiatric, 2013). This is a high percentage of males, but this is not surprising due to the history of this country. In the history of America, we used to live in what would be considered a "man's world" where men were seen to be superior to women. The norm of society was that men were the providers of the family, made all of the major decisions and were the head of the household. Women on the other hand, stayed at home raising the children and cooking meals.

In my clinical opinion, I believe that the reason that there is a higher percentage of males being diagnosed with Narcissistic Personality Disorder is because girls are taught to be nurturing and selfless human beings because of possibly one day using those skills for motherhood. Jennifer Goodman, in an article titled "Narcissistic Men Typically Direct Their Rage toward Woman," suggests that narcissistic males have a problem with rejection and people challenging them rather than people who are different (Goodwin, 2010). In other words, a Narcissistic male and an independent woman with common sense would not work well since that would bring a lot of conflict because you have a man who is convinced he is superior and more important than others, and a woman who has no narcissistic traits will see right through that behavior causing conflict by challenging his opinions. A Narcissistic individual's anger may be triggered by feelings of rejection and unacceptance.

In a 2010 study conducted by Scott Keller, 104 male college students were surveyed with questionnaires designed to measure narcissism. The survey asked questions like "I love to be in the center of attention or I am embarrassed to be in the center of attention." The men who scored the highest in narcissism viewed woman as "conniving gold diggers, who tempt men with sex and don't deliver" (Keiller, 2010).

Charles Darwinwas an English naturalist and philosopher who contributed to the development of evolutionary theory (Britannica, 2014). Using his ideas explaining narcissism in a cultural way, we can comparing statistics between African Americans and Europeans. For example, what is the percentage of narcissistic disorders diagnosed in minority communities and in European communities? In this day and age, an individual may assume that the number of Europeans diagnosed with Narcissistic Personality Disorder would be higher than minorities because there are people in society who believe in "white privilege." According to Tim Wise (2014), white privilege is any advantage, opportunity, benefit, head start, or general protection from negative social mistreatment which persons who are typically white would enjoy but generally others would not (Wise, 2014). I believe white privilege has been created by society, and it is an individual experience. White privilege is a term for societal privileges that benefit white people beyond what is commonly experienced by non-white people under the same social, political or economic circumstances.

"Darwin was, after all, a man of his time, class and society. True, he was committed to a monogenic, rather than the prevailing polygenic, view of human origins, but he still divided humanity into distinct races according to differences in skin, eye or hair color. He was also convinced that evolution was progressive, and that the white races—especially the Europeans—were evolutionarily more advanced than the black races, thus establishing race differences and a racial hierarchy. Darwin's views on gender, too, were utterly conventional. He stated that the result of sexual selection is for men to be, "more courageous, pugnacious and energetic than woman [with] a more inventive genius." (Rose, 2009)

White narcissism is based on an ideology that was born during slavery, namely, if you are privileged enough to own another human being then you begin to feel that you are truly special. According to the article "White Narcissism" by R. McDonald (2014), slave ownership was a breeding ground for arrogance, grandiosity, and entitlement. Both being a slave and slave owner is a psychological tragedy. Slave owners made sure that slaves did not spoil their children so that they would not have a sense of entitlement or equality. (McDonald, 2014)

"Evolutionary theorist Charles Darwin argued that there is only one race of humans, he noted in 1871 that others had attempted to identify up to sixty-three distinct races. The fact that scholars could not agree on racial distinctions, creating their own categories of race separate from each other, reveals that race is not an obvious biological fact, but is, instead, socially constructed. Many anthropologists historically classified humans into multiple races, but the majority of modern anthropologists argue, like Darwin, that there is only one race of human beings." (Halley, Eshldman, & Vijaya, 2011)

There was time where it was survival of the fittest, because only one social group could have the wealth and power. Historically, Europeans were a symbol of success and wealth, especially while slave trade and trade in crops is what they received profit from. My hypothesis is that among the high percentage in men that are diagnosed with Narcissistic Personality Disorder most of them are European decent because of being born in white privilege and being a White man in America.

However, after doing further research my hypothesis proved to be wrong. In 2008 Psychiatrist D. Dawson and his colleagues performed a face-to-face survey in California with 35,653 adults to present nationally representative findings on the prevalence, socio-demographics, and comorbidity of narcissistic personality disorder among men and women. The results showed that not only were the percentages higher in males (7.7%) than females (4.8%) but were also significantly more prevalent in the Black culture and non-married individuals (Dawson, et al., 2008).

If you look at both studies presented, you can conclude that Narcissistic Personality Disorder is more prevalent in the African American race, males and college students. Although I was proven wrong, I am not very shocked. In today's society, we have created the phrase, a credit to your race. According to Roush (2008) who wrote the book : "A Rational Approach to Race Relations", explains this phrase quite well, he states that people of all races who try to dispel racial stereotypes by modeling opposing behavior are admirable (Roush, 2008). One stereotype of black people is that they are lazy and do not pursue high education. A black man may be seen as having narcissistic behavior if he strives to attend college or graduate school. He may be seen for being narcissistic because he is seen as powerful among his race from other races. Although being a credit to your race is indeed insulting because it implies a negative cogitation to that race, it definitely gives that individual the recognition, and praise that can contribute to Narcissistic behavior in adulthood. I would like to see further research into the statistics behind Narcissism and other cultures. The DSM 5 should have more statistics on Narcissistic Personality Disorder pertaining to population.

References

Association, A. P. (2013). Diagnostic And Statistical Manual of Mental Disorders Fifth Edition . Arlington: VA.

Bannister, R. C. (2000). Social Darwinism . Retrieved from http://autocww.colorado.edu/~flc/E64ContentFiles/SociologyAndReform/Socia lDarwinism.html

Bender, D. (2014, 11 26). Data from Texas A&M University Advance Knowledge in Personality Disorders (Narcissistic Personality Disorder in DSM-5). Retrieved from http://www.4-traders.com/news/Data-from-Texas-AM-University-Advance-Knowledge-in-Personality-Disorders-Narcissistic-Personality--19457711/

Britannica. (2014, October). Social Darwinism. Retrieved from http://www.britannica.com/EBchecked/topic/551058/social-Darwinism

Dawson, D., Goldstein, R., Chou, S., Stinson, F., Huang, B., Smith, S. . . . Grant, B. (2008, July 7). Prevalence, correlates, disability, and comorbidity of DSM-IV narcissistic personality disorder: results from the wave 2 national epidemiologic survey on alcohol and related conditions. J Clin Psychiatry, pp. 1033-45.

Goodwin, J. (2010, August 6). Narcissistic Men typically direct their rage towards woman. Retrieved from Medicinenet.com: medicinenet.com/script/main/art.asp?articlekey=118736

Halley, J., Eshldman, A., & Vijaya, R. M. (2011). Seeing White: Introduction to White Privilege and Race. Rowman & Littlefield Publishers.

Keiller, S. (2010, July 23). Dsm-5 sex roles. Narcissistic Personality Disorder. New Philadelphia, Ohio.

McDonald, R. (2014, September 01). Retrieved from Friends Journal: www.friendsjournal.com/white-narcissism

Rose, S. (2009). Darwin, race and gender. Embo reports, 291-298.

Roush, R. V. (2008). Solutions to Race Resentment. In R. Roush, A Rational Approach To Race Relations (p. 219). iUniverse, incoporated .

Wise, T. (2014, May). White Privilege. Retrieved from http://www.timwise.org/f-a-q-s/

www.ingramcontent.com/pod-product-compliance
Lightning Source LLC
Chambersburg PA
CBHW032009170526
45157CB00002B/607